PRIMATES
AND
PHILOSOPHERS

PRIMATES
AND
PHILOSOPHERS

How Morality Evolved

Frans de Waal

Robert Wright
Christine M. Korsgaard
Philip Kitcher
Peter Singer

EDITED AND INTRODUCED BY
Stephen Macedo and Josiah Ober

PRINCETON UNIVERSITY PRESS

PRINCETON AND OXFORD

Published by Princeton University Press, 41 William Street, Princeton,
New Jersey 08540

In the United Kingdom: Princeton University Press, 6 Oxford Street, Woodstock,
Oxfordshire OX20 1TW

Fifth printing, and first paperback printing, 2009
Paperback ISBN 978-0-691-14129-9

The Library of Congress has cataloged the cloth edition of this book as follows

Waal, F. B. M. de (Frans B. M.), 1948–
Primates and philosophers : how morality evolved / Frans de Waal ; edited and
introduced by Stephen Macedo and Josiah Ober ; Christine M. Korsgaard . . . [et al.].
p. cm.
Includes bibliographical references and index.
ISBN-13: 978-0-691-12447-6 (hardcover : alk. paper)
ISBN-10: 0-691-12447-7 (hardcover : alk. paper)
1. Ethics, Evolutionary. 2. Primates—Behavior. 3. Altruistic behavior in
animals. I. Macedo, Stephen, 1957– II. Ober, Josiah. III. Korsgaard,
Christine M. (Christine Marion) IV. Title.
BJ1311.W14 2006
171'.7—dc22
2006013905

British Library Cataloging-in-Publication Data is available
This book has been composed in Minion Family & Minion Condensed
Printed on acid-free paper. ∞
press.princeton.edu

Printed in the United States of America

7 9 10 8

Contents

𝄢

PART III Response to Commentators

Acknowledgments

𝒟

I would like to thank Philip Kitcher, Christine M. Korsgaard, Richard Wrangham, and Robert A. Wright, who were the commentators for the Tanner Lectures that I gave at Princeton University in November of 2003. Also, I thank Peter Singer for his comments, which appear in this publication, and Stephen Macedo and Josiah Ober for their introduction. I am grateful to the Tanner Foundation, which endows the Tanner Lecture Series; to the Princeton University Press, with special thanks to Sam Elworthy and Jodi Beder, editor and copy editor; and to the staff of the Center for Human Values who organized the lectures and helped coordinate this book's production: Stephen Macedo, director; Will Gallaher, former associate director; and Jan Logan, assistant director. Finally, I am grateful to Emory University's Yerkes National Primate Research Center, in Atlanta, Georgia, and other centers and zoos at which I have conducted research, as well as to all of my many collaborators and graduate students for helping collect the data presented here.

Frans de Waal
March 2006

Introduction

JOSIAH OBER AND STEPHEN MACEDO

In the **Tanner Lectures** on Human Values that became the lead essay in this book, Frans de Waal brings his decades of work with primates, and his habit of thinking deeply about the meaning of evolution, to bear upon a fundamental question about human morality. Three distinguished philosophers and a prominent student of evolutionary psychology then respond to the way de Waal's question is framed, and to his answer. Their essays are at once appreciative of de Waal's endeavor and critical of certain of his conclusions. De Waal responds to his critics in an afterword. While there is considerable disagreement among the five essayists about both the question and how to answer it, they also share a good deal of common ground. First, all contributors to this book accept the standard scientific account of biological evolution as based on random natural selection. None suggests that there is any reason to suppose that humans are different in their metaphysical essence from other animals, or at least, none base their arguments on the idea that humans uniquely possess a transcendent soul.

x JOSIAH OBER AND STEPHEN MACEDO

A second important premise that is shared by de Waal and all four of his commentators is that moral goodness is something real, about which it is possible to make truth claims. Goodness requires, at a minimum, taking proper account of others. Badness, by the same token, includes the sort of selfishness that leads us to treat others improperly by ignoring their interests or treating them as mere instruments. The two basic premises of evolutionary science and moral reality establish the boundaries of the debate over the origins of goodness as it is set forth in this book. This means that those religious believers who are committed to the idea that humans have been uniquely endowed with special attributes (including a moral sense) by divine grace alone are not participants in the discussion as it is presented here. Nor are social scientists committed to a version of rational agent theory that regards the essence of human nature as an irreducible tendency to choose selfishness (free-riding, cheating) over voluntary cooperation. Nor, finally, are moral relativists, who believe that an action can be judged as right or wrong only locally, by reference to contingent and contextual considerations. So what we offer in this volume is a debate among five scholars who agree on some basic issues about science and morality. It is a serious and lively conversation among a group of thinkers who are deeply committed to the value and validity of science and to the value and reality of other-regarding morality.

The question that de Waal and his commentators seek to address is this: How, given that there are strong scientific reasons to suppose that selfishness (at least at the genetic level) is a primary mechanism of natural selection, did we humans come to be so strongly attached to the value of goodness? Or, to put it a bit differently, why don't we think it is good to be bad? For those who believe that morality is real, but that it cannot be explained or justified simply by resort to the theological assumption that a unique human propensity to goodness is a product of a divine grace, this is a hard problem, and an important one.

De Waal's aim is to argue against a set of answers to his "whence morality?" question that he describes as "Veneer Theory"—the argument that morality is only a thin veneer overlaid on an amoral or immoral core. De Waal suggests that Veneer Theory is (or at least was until recently) quite widely held. His primary target is Thomas Huxley, a scientist dubbed "Darwin's bulldog" for his fierce defense of Darwin's theory of evolution against its late-nineteenth-century detractors. De Waal argues that Huxley betrayed his own core Darwinian commitments in advocating a view of morality as "garden tending"—a constant battle against the luxuriant weeds of immoralism that perennially threaten to take over the human psyche. De Waal's other targets include some social contract theorists (notably Thomas Hobbes) who begin with a conception of humans as fundamentally asocial or even antisocial, and some evolutionary biologists who, in his view, tend to overgeneralize from the established role of selfishness in the natural selection process.

None of the five essayists in this volume identifies him- or herself as a "veneer theorist" in de Waal's sense. Yet, as the essays show, Veneer Theory can be conceived in various ways. It may therefore be useful to describe a sort of ideal type of VT, even if this risks setting up a straw man. Ideal-type Veneer Theory assumes that humans are by nature bestial and therefore bad—that is, narrowly selfish—and thus should be expected to act badly—that is, to treat others improperly. Yet it is an observable fact that at least sometimes humans treat each other well and properly, just as if we were good. Since, by the argument, humans are basically bad, their good behavior must be explained as the product of a veneer of morality, mysteriously laid over the bad natural core. De Waal's primary objection is that VT cannot identify the *source* of this veneer of goodness. The veneer is something that apparently exists outside nature and so must be rejected as a myth by anyone committed to scientific explanation of natural phenomena.

If the Veneer Theory of moral goodness is based on a myth, the phenomenon of human goodness must be explained in some other way. De Waal begins by reversing the initial premise: Humans are, he suggests, by nature good. Our "good nature" is inherited, along with much else, from our nonhuman ancestors through the ordinary Darwinian process of natural selection. In order to test this premise, he invites us to join him in looking carefully at the behavior our closest nonhuman relatives—first at chimpanzees, and then at other primates more distantly related to ourselves, and ultimately at non-primate social animals. If our closest relatives do in fact act as if they were good, and if we humans also act as if we were good, the methodological principle of parsimony urges us to suppose that the goodness is real, that the motivation for goodness is natural, and that the morality of humans and their relatives has a common source.

While human behavioral goodness is more fully developed than nonhuman behavioral goodness, the simpler nonhuman morality must, according to de Waal, be regarded, in a substantial sense, as the *foundation* of more complex human morality. The empirical evidence for de Waal's "anti-veneer" theory linking human and nonhuman morality consists of careful observations of the behavior of humanity's relatives.

De Waal has spent a long and extremely successful career minutely observing primate behavior and he has seen and recorded much goodness. In the process he has developed immense respect and fondness for his subjects. One part of the pleasure of reading de Waal on primates, a pleasure that radiates in each of the commentators' essays, is his evident joy in his years of working with chimpanzees, bonobos, and capuchins, and his sense of them as his collaborators in a momentous undertaking.

De Waal concludes that the human capacity to act well at least sometimes, rather than badly all the time, has its evolutionary origins in emotions that we share with other animals—in involuntary (unchosen, pre-rational) and physiologically obvious (thus observable) responses to the circumstances of others. A fundamentally important form of emotional response is empathy. De Waal explains that the empathetic reaction is, in the first instance, a matter of "emotional contagion." Creature A identifies directly with the circumstances of creature B, coming, as it were, to "feel his or her pain." At this level, empathy is still in a sense selfish—A seeks to comfort B because A has "caught" B's pain and is himself seeking comfort. At a more advanced level, however, emotional empathy can yield sympathy—that is, the recognition that B has situationally specific wants or needs that are different from those of A. De Waal offers the lovely and telling example of a chimpanzee trying to help an injured bird to fly away. Since flying is an action the chimpanzee herself obviously could never perform, the ape is responding to the bird's particular needs and its distinctive way of being in the world.

Emotional contagion is commonly observed in many species; sympathy is only observed among certain of the great apes. Related emotional responses conducive to good behavior include reciprocal altruism and perhaps even a sense of fairness—although this last remains disputed (as Philip Kitcher points out). Once again, the most complex and sophisticated forms of these emotion-motivated (as de Waal argues) behaviors are uniquely observed among apes and a few other species—elephants, dolphins, and capuchins.

Emotional responses are, de Waal argues, the "building blocks" of human morality. Human moral behavior is considerably more elaborate than that of any nonhuman animal, but, in de Waal's view, it is *continuous with* nonhuman behavior—just as sympathy in chimpanzees is more elaborate than but continuous with emotional contagion in other animals. Given this continuity of good nature, there is no need to imagine morality being mysteriously added to an immoral core. De Waal invites us to imagine ourselves, not as solid clay garden trolls covered by a thin veneer of gaudy paint, but as "Russian dolls"—our external moral selves are ontologically continuous with a nested series of inner "prehuman selves." And all the way down to the tiny little figure in the very center, these selves are homogeneously "goodnatured."

As the vigor of the four responses demonstrates, de Waal's conception of the origins and nature of human morality is a challenging one. Each of the commentators agrees with de Waal that ideal-type Veneer Theory is unattractive on the face of it, although they disagree on exactly what VT is, or whether any reasonable person could subscribe to it, at least in the robust form sketched above. Yet at the end of the day each of the commentators has developed something that might be described as a distant cousin of Veneer Theory. Robert Wright is forthright about this, calling his position "naturalistic Veneer Theory." Indeed, as Peter Singer points out (p. 145), de Waal himself at one point speaks of how "fragile" is the human effort to expand the "circle of morality" to outsiders—a locution that seems to invite imagining at least certain extended forms of human morality as a sort of veneer.

De Waal's concern for how far the "circle of morality" can be expanded without becoming untenably fragile underlines the issue that leads his commentators to draw a bright line between human morality and animal behavior. This is their firm conviction that "genuine" (Kitcher) morality must also be universalizable. This conviction excludes animals from the ambit of genuinely moral beings. It places them "beyond moral judgment," in Korsgaard's words, because nonhuman animals do not universalize their good behavior. The tendency towards partiality for insiders is a constant among nonhuman social animals. Admittedly, the same partial tendency may be endogenous to humans, as de Waal believes. And it may be an endemic threat to human morality, as Robert Wright argues. But, as Kitcher, Korsgaard, and Singer all point out, the universalization of the set of beings (all persons, or, with Singer, all creatures with interests) to which moral duties are owed is treated as *conceptually feasible* by humans (and as conceptually essential by some human philosophers). And it is at least sometimes put into practice by them.

Each commentator asks a similar question, albeit in quite different philosophical registers: If even the most advanced nonhuman animals ordinarily limit their good behavior to insiders (kin or community members), can we really speak of their behavior as *moral?* And if the answer is no (as each concludes), then we must assume that human beings have some capacity that is *discontinuous* with the natural capacities of all nonhuman species. De Waal acknowledges the issue, noting (as Singer again points out, p. 144) that "It is *only* when we make general, impartial judgments that we can really begin to speak of *moral* approval and disapproval."

The most obvious capacity discontinuity between humans and nonhuman animals is in the area of speech, and the self-conscious employment of reason that we associate strongly with the uniquely human use of language. Speech, language use, and reason are obviously connected to cognition. So what can we say about nonhuman cognition? No one participating in this collection supposes that any nonhuman species is the cognitive *equal* of human beings, but the question remains whether humans are *uniquely* capable of moral reasoning.

This is the point in the debate at which defining *anthropomorphism* becomes a lively issue; Wright in particular focuses on the importance of the anthropomorphism question. De Waal is an ardent and thoughtful advocate of a critical and parsimonious version of scientific anthropomorphism— which he sharply distinguishes from the scattershot sentimental anthropomorphism typical of much (albeit delightful) popular writing about animals. None of the four commentators can fairly be characterized as an advocate of "anthropodenial"—de Waal's term for the practice of those who, perhaps out of an aesthetic horror of nature, refuse to acknowledge the continuities between humans and other animals. Much of the debate among philosophers and animal behaviorists over human uniqueness has centered on the question of whether any nonhuman animal is capable of developing anything like a real Theory of Mind (ToM)—that is to say, whether or not the capacity to imagine the contents of another being's mind as different from one's own is uniquely human. There is some experimental data that may lend support to both sides of this question. De Waal answers doubters by noting that individual chimpanzees can recognize themselves in a mirror (thus demonstrating the self-consciousness often supposed as an antecedent condition for ToM). He pointedly draws our attention to the stark anthropocentrism of demanding that ape subjects be able to formulate a theory of *human* minds. But the question of nonhuman ToM remains undecided; clearly more research on this area is called for.

Kitcher and Korsgaard sharply distinguish animal behavior motivated by emotion from human morality, which they argue must be based on cognitive self-consciousness about the propriety of one's proposed line of action. Kitcher draws the line by making Humean/Smithian "spectatorism" into a kind of self-consciousness requiring speech. Korsgaard appeals to the Kantian conception of autonomous self-governance as the necessary foundation for genuine morality. Both Kitcher and Korsgaard describe nonhuman animals as "wantons," helping themselves to a concept developed in other contexts by the moral philosopher Harry Frankfurt. Frankfurtian wantons lack a mechanism by which to discriminate in any consistent way among the various motivations that from time to time might prompt them to act. And thus wantons cannot be said to be guided by self-conscious reasoning on the propriety of their proposed actions. Yet at this point the question arises of whether Kitcher and Korsgaard are setting the bar of morality at a level that most *human* action fails to reach. Each philosopher offers a self-consciously *normative* account of morality as how people ought to act, rather than a *descriptive* account of how most of us actually do act most of the time. If most humans, in their actual behavior, act like wantons, it takes some of the sting out of the claim that all nonhuman animals act like wantons too.

The same issue arises in Singer's discussion of what moral philosophers call "trolley problems." Singer's consequentialist concern with aggregating interests leads him to claim that moral reason demands that, under the right circumstances, one ought to push another human being in front of a runaway trolley in order to save five others (the premise is that one's own body is too light to stop the trolley, whereas the pushed individual is of sufficient mass). Singer alludes to studies of brain scans of individuals as they answer the question of how one should act in the "kill one to save five" situation. People who say that one should *not* kill in that situation make quick judgments and their brain activity at the moment of decision is concentrated in areas associated with emotion. Those who say that one ought to kill manifest increased activity in parts of the brain associated with rational cognition. Singer thus claims that what he regards as the morally correct answer is also the cognitively rational answer. Yet Singer also acknowledges that those giving the correct answer are in the minority: *most* people do *not* say that they would choose to act personally to kill one individual to save five others. Singer does not cite any cases of people *actually* pushing others in front of trolleys.

The point is that de Waal's evidence, quantitative and anecdotal, for primate emotional response is based *entirely* on observations of actual behavior. De Waal must base his account of primate morality on how primates do in fact act because he has no access to their "ought" stories about what moral reason might ideally demand of them, or to how they suppose they ought to act in a hypothetical situation. So there seems to be a risk of comparing apples and oranges: contrasting primate *behavior* (based on quantitative and anecdotal observation) with human *normative ideals.* Of course, de Waal's critics can respond that the difference among comparanda is precisely the point: nonhuman animals have not *got* any ought stories, or for that matter stories of any kind, because they lack the capacity for speech, language, and reason.

Nonhuman animals cannot enunciate normative ideals, to one another or to us. Does that fact require us to draw a bright line between the kinds of emotion-motivated "moral" behavior that de Waal and others have observed in primates and the "genuine" reason-based moral actions of humans? If the copy editor of this book knew the right answer to that question he or she would know which word in the previous sentence—"moral" or "genuine"—should have its scare quotes struck out. Much in our understanding of ourselves, and the other species with which we share the earth, rests on that choice. One goal of this book is to encourage each reader to think carefully about how he or she would choose to wield the imagined editorial pencil—to invite each of you to attend to this and other conversations among the set of scholars who think hard and care passionately about primate behavior and the set of those who think hard and care equally passionately about human morality. The existence of this book is proof that those two sets are partially overlapping. Part of its purpose is to advocate an increase in the extent of the overlap and to promote thoughtful discussion among all those concerned with goodness and its origins, in human and nonhuman animals alike.

PART I

MORALLY EVOLVED

PRIMATE SOCIAL INSTINCTS, HUMAN MORALITY, AND THE RISE AND FALL OF "VENEER THEORY"

Frans de Waal

We approve and we disapprove because we cannot do
otherwise. Can we help feeling pain when the fire burns us?
Can we help sympathizing with our friends?
　—Edward Westermarck (1912 [1908]: 19)

Why should our nastiness be the baggage of an apish past
and our kindness uniquely human? Why should we not seek
continuity with other animals for our "noble" traits as well?
　—Stephen Jay Gould (1980: 261)

H *omo homini lupus*—"man is wolf to man"—is an
ancient Roman proverb popularized by Thomas
Hobbes. Even though its basic tenet permeates large
parts of law, economics, and political science, the proverb
contains two major flaws. First, it fails to do justice to canids,
which are among the most gregarious and cooperative animals on the planet (Schleidt and Shalter 2003). But even
worse, the saying denies the inherently social nature of our
own species.

Social contract theory, and Western civilization with it,
seems saturated with the assumption that we are asocial, even
nasty creatures rather than the *zoon politikon* that Aristotle
saw in us. Hobbes explicitly rejected the Aristotelian view by
proposing that our ancestors started out autonomous and
combative, establishing community life only when the cost of
strife became unbearable. According to Hobbes, social life

never came naturally to us. He saw it as a step we took reluctantly and "by covenant only, which is artificial" (Hobbes 1991 [1651]: 120). More recently, Rawls (1972) proposed a milder version of the same view, adding that humanity's move toward sociality hinged on conditions of fairness, that is, the prospect of mutually advantageous cooperation among equals.

These ideas about the origin of the well-ordered society remain popular even though the underlying assumption of a rational decision by inherently asocial creatures is untenable in light of what we know about the evolution of our species. Hobbes and Rawls create the illusion of human society as a voluntary arrangement with self-imposed rules assented to by free and equal agents. Yet, there never was a point at which we became social: descended from highly social ancestors—a long line of monkeys and apes—we have been group-living forever. Free and equal people never existed. Humans started out—if a starting point is discernible at all—as interdependent, bonded, and unequal. We come from a long lineage of hierarchical animals for which life in groups is not an option but a survival strategy. Any zoologist would classify our species as *obligatorily gregarious*.

Having companions offers immense advantages in locating food and avoiding predators (Wrangham 1980; van Schaik 1983). Inasmuch as group-oriented individuals leave more offspring than those less socially inclined (e.g., Silk et al. 2003), sociality has become ever more deeply ingrained in primate biology and psychology. If any decision to establish societies was made, therefore, credit should go to Mother Nature rather than to ourselves.

This is not to dismiss the heuristic value of Rawls's "original position" as a way of getting us to reflect on what kind of society we would *like* to live in. His original position refers to a "purely hypothetical situation characterized so as to lead to certain conceptions of justice" (Rawls 1972: 12). But even if we do not take the original position literally, hence adopt it only for the sake of argument, it still distracts from the more pertinent argument that we ought to be pursuing, which is how we actually came to be what we are today. Which parts of human nature have led us down this path, and how have these parts been shaped by evolution? Addressing a real rather than hypothetical past, such questions are bound to bring us closer to the truth, which is that we are social to the core.

A good illustration of the thoroughly social nature of our species is that, second to the death penalty, solitary confinement is the most extreme punishment we can think of. It works this way only, of course, because we are not born as loners. Our bodies and minds are not designed for life in the absence of others. We become hopelessly depressed without social support: our health deteriorates. In one recent experiment, healthy volunteers deliberately exposed to cold and flu viruses got sick more easily if they had fewer friends and family around (Cohen et al. 1997). While the primacy of connectedness is naturally understood by women—perhaps because mammalian females with caring tendencies have outreproduced those without for 180 million years—it applies equally to men. In modern society, there is no more effective way for men to expand their age horizon than to get and stay married: it increases their chance of living past the age of sixty-five from 65 to 90 percent (Taylor 2002).

Our social makeup is so obvious that there would be no need to belabor this point were it not for its conspicuous absence from origin stories within the disciplines of law, economics, and political science. A tendency in the West to see

6 FRANS DE WAAL

emotions as soft and social attachments as messy has made theoreticians turn to cognition as the preferred guide of human behavior. We celebrate rationality. This is so despite the fact that psychological research suggests the primacy of affect: that is, that human behavior derives above all from fast, automated emotional judgments, and only secondarily from slower conscious processes (e.g., Zajonc 1980, 1984; Bargh and Chartrand 1999).

Unfortunately, the emphasis on individual autonomy and rationality and a corresponding neglect of emotions and attachment are not restricted to the humanities and social sciences. Within evolutionary biology, too, some have embraced the notion that we are a self-invented species. A parallel debate pitting reason against emotion has been raging regarding the origin of morality, a hallmark of human society. One school views morality as a cultural innovation achieved by our species alone. This school does not see moral tendencies as part and parcel of human nature. Our ancestors, it claims, became moral by choice. The second school, in contrast, views morality as a direct outgrowth of the social instincts that we share with other animals. In the latter view, morality is neither unique to us nor a conscious decision taken at a specific point in time: it is the product of social evolution.

The first standpoint assumes that deep down we are not truly moral. It views morality as a cultural overlay, a thin veneer hiding an otherwise selfish and brutish nature. Until recently, this was the dominant approach to morality within evolutionary biology as well as among science writers popularizing this field. I will use the term "Veneer Theory" to denote these ideas, tracing their origin to Thomas Henry Huxley (although they obviously go back much further in Western philosophy and religion, all the way to the concept of original sin). After treating these ideas, I review Charles Darwin's quite different standpoint of an evolved morality, which was inspired by the Scottish Enlightenment. I further discuss the views of Mencius and Westermarck, which agree with those of Darwin.

Given these contrasting opinions about continuity versus discontinuity with other animals, I then build upon an earlier treatise (de Waal 1996) in paying special attention to the behavior of nonhuman primates in order to explain why I think the building blocks of morality are evolutionarily ancient.

VENEER THEORY

In 1893, for a large audience in Oxford, England, Huxley publicly reconciled his dim view of the natural world with the kindness occasionally encountered in human society. Huxley realized that the laws of the physical world are unalterable. He felt, however, that their impact on human existence could be softened and modified if people kept nature under control. Thus, Huxley compared humanity with a gardener who has a hard time keeping weeds out of his garden. He saw human ethics as a victory over an unruly and nasty evolutionary process (Huxley 1989 [1894]).

This was an astounding position for two reasons. First, it deliberately curbed the explanatory power of evolution. Since many consider morality the essence of humanity, Huxley was in effect saying that what makes us human could not be handled by evolutionary theory. We can become moral only by opposing our own nature. This was an inexplicable retreat by someone who had gained a reputation as "Darwin's Bulldog" owing to his fierce advocacy of evolution. Second,

Huxley gave no hint whatsoever where humanity might have unearthed the will and strength to defeat the forces of its own nature. If we are indeed born competitors, who don't care about the feelings of others, how did we decide to transform ourselves into model citizens? Can people for generations maintain behavior that is out of character, like a shoal of piranhas that decides to turn vegetarian? How deep does such a change go? Would not this make us wolves in sheep's clothing: nice on the outside, nasty on the inside?

This was the only time Huxley broke with Darwin. As Huxley's biographer, Adrian Desmond (1994: 599), put it: "Huxley was forcing his ethical Ark against the Darwinian current which had brought him so far." Two decades earlier, in *The Descent of Man*, Darwin (1982 [1871]) had unequivocally included morality in human nature. The reason for Huxley's departure has been sought in his suffering at the cruel hand of nature, which had taken the life of his beloved daughter, as well as his need to make the ruthlessness of the Darwinian cosmos palatable to the general public. He had depicted nature as so thoroughly "red in tooth and claw" that he could maintain this position only by dislodging human ethics, presenting it as a separate innovation (Desmond 1994). In short, Huxley had talked himself into a corner.

Huxley's curious dualism, which pits morality against nature and humanity against other animals, was to receive a respectability boost from Sigmund Freud's writings, which throve on contrasts between the conscious and subconscious, the ego and superego, Love and Death, and so on. As with Huxley's gardener and garden, Freud was not just dividing the world into symmetrical halves: he saw struggle everywhere. He explained the incest taboo and other moral restrictions as the result of a violent break with the freewheeling sexual life of the primal horde, culminating in the collective slaughter of an overbearing father by his sons (Freud 1962 [1913]). He let civilization arise out of the renunciation of instinct, the gaining of control over the forces of nature, and the building of a cultural superego (Freud 1961 [1930]).

Humanity's heroic combat against forces that try to drag him down remains a dominant theme within biology today, as illustrated by quotes from outspoken Huxleyans. Declaring ethics a radical break with biology, Williams wrote about the wretchedness of nature, culminating in his claim that human morality is a mere by-product of the evolutionary process: "I account for morality as an accidental capability produced, in its boundless stupidity, by a biological process that is normally opposed to the expression of such a capability" (Williams 1988: 438).

Having explained at length that our genes know what is best for us, programming every little wheel of the human survival machine, Dawkins waited until the very last sentence of *The Selfish Gene* to reassure us that, in fact, we are welcome to chuck all of those genes out the window: "We, alone on earth, can rebel against the tyranny of the selfish replicators" (Dawkins 1976: 215). The break with nature is obvious in this statement, as is the uniqueness of our species. More recently, Dawkins (1996) has declared us "nicer than is good for our selfish genes," and explicitly endorsed Huxley: "What I am saying, along with many other people, among them T. H. Huxley, is that in our political and social life we are entitled to throw out Darwinism, to say we don't want to live in a Darwinian world" (Roes, 1997: 3; also Dawkins 2003).

Darwin must be turning in his grave, because the implied "Darwinian world" is miles removed from what he himself

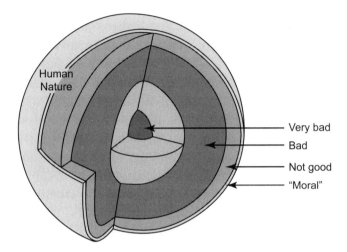

Figure 1 The popular view of morality among biologists during the past quarter of a century was summarized by Ghiselin (1974: 247): "Scratch an 'altruist,' and watch a 'hypocrite' bleed." Humans were considered thoroughly selfish and competitive, with morality being no more than an afterthought. Summarized as "Veneer Theory," this idea goes back to Darwin's contemporary, Thomas Henry Huxley. It is visualized here tongue-in-cheek as human nature bad to its core.

envisioned (see below). What is lacking in these statements is any indication of how we can possibly negate our genes, which the same authors at other times have depicted as all-powerful. Like the views of Hobbes, Huxley, and Freud, the thinking is thoroughly dualistic: we are part nature, part culture, rather than a well-integrated whole. Human morality is presented as a thin crust underneath of which boil antisocial, amoral, and egoistic passions. This view of morality as a veneer was best summarized by Ghiselin's famous quip: "Scratch an 'altruist,' and watch a 'hypocrite' bleed" (Ghiselin 1974: 247; figure 1).

Veneer Theory has since been popularized by countless science writers, such as Wright (1994), who went so far as to claim that virtue is absent from people's hearts and souls, and that our species is potentially but not naturally moral. One might ask: "But what about the people who occasionally experience in themselves and others a degree of sympathy, goodness, and generosity?" Echoing Ghiselin, Wright replies that the "moral animal" is essentially a hypocrite:

> [T]he pretense of selflessness is about as much part of human nature as is its frequent absence. We dress ourselves up in tony moral language, denying base motives and stressing our at least minimal consideration for the greater good; and we fiercely and self-righteously decry selfishness in others. (Wright 1994: 344)

To explain how we manage to live with ourselves despite this travesty, theorists have called upon self-deception. If people think they are at times unselfish, so the argument goes, they must be hiding their true motives from themselves (e.g., Badcock 1986). In the ultimate twist of irony, anyone who fails to believe that we are fooling ourselves, and feels that genuine kindness actually exists in the world, is considered a wishful thinker, hence accused of fooling him- or herself.

Some scientists have objected, however:

> It is frequently said that people endorse such hypotheses [about human altruism] because they *want* the world to be a friendly and hospitable place. The defenders of egoism and individualism who advance this criticism thereby pay themselves a compliment; they pat themselves on the back for

staring reality squarely in the face. Egoists and individualists
are objective, they suggest, whereas proponents of altruism
and group selection are trapped by a comforting illusion.
(Sober and Wilson 1998: 8–9)

These back-and-forth arguments about how to reconcile
everyday human kindness with evolutionary theory seem an
unfortunate legacy of Huxley, who had a poor understanding of the theory that he so effectively defended against its
detractors. In the words of Mayr (1997: 250): "Huxley, who
believed in final causes, rejected natural selection and did
not represent genuine Darwinian thought in any way. . . . It
is unfortunate, considering how confused Huxley was, that
his essay [on ethics] is often referred to even today as if it
were authoritative."

It should be pointed out, though, that in Huxley's time
there was already fierce opposition to his ideas (Desmond
1994), some of which came from Russian biologists, such as
Petr Kropotkin. Given the harsh climate of Siberia, Russian
scientists traditionally were far more impressed by the battle of animals against the elements than against each other,
resulting in an emphasis on cooperation and solidarity that
contrasted with Huxley's dog-eat-dog perspective (Todes
1989). Kropotkin's (1972 [1902]) *Mutual Aid* was an attack
on Huxley, but written with great deference for Darwin.

Although Kropotkin never formulated his theory with the
precision and evolutionary logic available to Trivers (1971)
in his seminal paper on reciprocal altruism, both pondered
the origins of a cooperative, and ultimately moral, society
without invoking false pretense, Freudian denial schemes, or
cultural indoctrination. In this they proved the true followers of Darwin.

DARWIN ON ETHICS

Evolution favors animals that assist each other if by doing so they achieve long-term benefits of greater value than the benefits derived from going it alone and competing with others. Unlike cooperation resting on simultaneous benefits to all parties involved (known as mutualism), reciprocity involves exchanged acts that, while beneficial to the recipient, are costly to the performer (Dugatkin 1997). This cost, which is generated because there is a time lag between giving and receiving, is eliminated as soon as a favor of equal value is returned to the performer (for treatments of this issue since Trivers 1971, see Axelrod and Hamilton 1981; Rothstein and Pierotti 1988; Taylor and McGuire 1988). It is in these theories that we find the germ of an evolutionary explanation of morality that escaped Huxley.

It is important to clarify that these theories do not conflict by any means with popular ideas about the role of selfishness in evolution. It is only recently that the concept of "selfishness" has been plucked from the English language, robbed of its vernacular meaning, and applied outside of the psychological domain. Even though the term is seen by some as synonymous with self-serving, English does have different terms for a reason. Selfishness implies the *intention* to serve oneself, hence knowledge of what one stands to gain from a particular behavior. A vine may be self-serving by overgrowing and suffocating a tree; but since plants lack intentions, they cannot be selfish except in a meaningless, metaphorical sense. Unfortunately, in complete violation of the term's original meaning, it is precisely this empty sense of "selfish" that has come to dominate debates about human nature. If our genes are selfish, we must be selfish, too, is the argument one often hears, despite the fact that genes are mere molecules, and hence cannot be selfish (Midgley 1979).

It is fine to describe animals (and humans) as the product of evolutionary forces that promote self-interests so long as one realizes that this by no means precludes the evolution of altruistic and sympathetic tendencies. Darwin fully recognized this, explaining the evolution of these tendencies by group selection instead of the individual and kin selection favored by modern theoreticians (but see, e.g., Sober and Wilson 1998; Boehm 1999). Darwin firmly believed his theory capable of accommodating the origins of morality and did not see any conflict between the harshness of the evolutionary process and the gentleness of some of its products. Rather than presenting the human species as falling outside of the laws of biology, Darwin emphasized continuity with animals even in the moral domain:

> Any animal whatever, endowed with well-marked social instincts, the parental and filial affections being here included, would inevitably acquire a moral sense or conscience, as soon as its intellectual powers had become as well developed, or nearly as well developed, as in man. (Darwin 1982 [1871]: 71–72)

It is important to dwell on the capacity for sympathy hinted at here and expressed more clearly by Darwin elsewhere (e.g., "Many animals certainly sympathize with each other's distress or danger" [Darwin 1982 (1871): 77]), because it is in this domain that striking continuities exist between humans and other social animals. To be vicariously affected by the emotions of others must be very basic, because these reactions have been reported for a great variety of animals and are often immediate and uncontrollable. They probably first emerged with parental care, in which vulnerable individuals are fed and protected. In many animals they stretch beyond this domain, however, to relations among unrelated adults (section 4 below).

In his view of sympathy, Darwin was inspired by Adam Smith, the Scottish moral philosopher and father of economics. It says a great deal about the distinctions we need to make between self-serving behavior and selfish motives that Smith, best known for his emphasis on self-interest as the guiding principle of economics, also wrote about the universal human capacity of sympathy:

> How selfish soever man may be supposed, there are evidently some principles in his nature, which interest him in the fortune of others, and render their happiness necessary to him, though he derives nothing from it, except the pleasure of seeing it. (Smith 1937 [1759]: 9)

The evolutionary origin of this inclination is no mystery. All species that rely on cooperation—from elephants to wolves and people—show group loyalty and helping tendencies. These tendencies evolved in the context of a close-knit social life in which they benefited relatives and companions able to repay the favor. The impulse to help was therefore never totally without survival value to the ones showing the impulse. But, as so often, the impulse became divorced from the consequences that shaped its evolution. This permitted its expression even when payoffs were unlikely, such as when strangers were beneficiaries. This brings animal altruism much closer to that of humans than usually thought, and explains the call for the temporary removal of ethics from the hands of philosophers (Wilson 1975: 562).

Personally, I remain unconvinced that we need group selection to explain the origin of these tendencies—we seem to get quite far with the theories of kin selection and reciprocal altruism. Moreover, there is so much intergroup migration (hence gene flow) in nonhuman primates that the conditions for group selection do not seem fulfilled. In all of the primates, the younger generation of one sex or another (males in many monkeys, females in chimpanzees and bonobos) tends to leave the group to join neighboring groups (Pusey and Packer 1987). This means that primate groups are far from genetically isolated, which makes group selection unlikely.

In discussing what constitutes morality, the actual behavior is less important than the underlying capacities. For example, instead of arguing that food-sharing is a building block of morality, it is rather the capacities thought to underlie food-sharing (e.g., high levels of tolerance, sensitivity to others' needs, reciprocal exchange) that seem relevant. Ants, too, share food, but likely based on quite different urges than those that make chimpanzees or people share (de Waal 1989a). This distinction was understood by Darwin, who looked beyond the actual behavior at the underlying emotions, intentions, and capacities. In other words, whether animals are nice to each other is not the issue, nor does it matter much whether their behavior fits our moral preferences or not. The relevant question rather is whether they possess capacities for reciprocity and revenge, for the enforcement of social rules, for the settlement of disputes, and for sympathy and empathy (Flack and de Waal 2000).

This also means that calls to reject Darwinism in our daily lives so as to build a moral society are based on a profound misreading of Darwin. Since Darwin saw morality as an evolutionary product, he envisioned an eminently more livable world than the one proposed by Huxley and his followers, who believe in a culturally imposed, artificial morality that receives no helping hand from human nature. Huxley's world is by far the colder, more terrifying place.

EDWARD WESTERMARCK

Edward Westermarck, a Swedish Finn who lived from 1862 until 1939, deserves a central position in any debate about the origin of morality, since he was the first scholar to promote an integrated view including both humans and animals and both culture and evolution. That his ideas were underappreciated during his lifetime is understandable, because they flew in the face of the Western dualistic tradition that pits body against mind and culture against instinct.

Westermarck's books are a curious blend of dry theorizing, detailed anthropology, and secondhand animal stories. The author was eager to connect human and animal behavior, but his own work focused entirely on people. Since at the time little systematic research on animal behavior existed, he had to rely on anecdotes, such as the one of a vengeful camel that had been excessively beaten on multiple occasions by a fourteen-year-old camel driver for loitering or turning the wrong way. The camel passively took the punishment; but a few days later, finding itself unladen alone on the road with the same driver, "seized the unlucky boy's head in its monstrous mouth, and lifting him up in the air flung him down again on the earth with the upper part of the skull completely torn off, and his brains scattered on the ground" (Westermarck 1912 [1908]: 38).

We should not discard such unverified reports out of

hand: stories of delayed retaliation abound in the zoo world, especially about apes and elephants. We now have systematic data on how chimpanzees punish negative actions with other negative actions (called a "revenge system" by de Waal and Luttrell 1988), and how a macaque attacked by a dominant member of its troop will turn around to redirect aggression against a vulnerable younger relative of its attacker (Aureli et al. 1992). These reactions fall under Westermarck's retributive emotions, but for him the term "retributive" went beyond its usual connotation of getting even. It also covered positive emotions, such as gratitude and the repayment of services. Depicting the retributive emotions as the cornerstone of morality, Westermarck weighed in on the question of its origin while anticipating modern discussions of evolutionary ethics.

Westermarck is part of a long tradition, going back to Aristotle and Thomas Aquinas, which firmly anchors morality in the natural inclinations and desires of our species (Arnhart 1998, 1999). Emotions occupy a central role; it is well known that, rather than being the antithesis of rationality, emotions aid human reasoning. People can reason and deliberate as much as they want, but, as neuroscientists have found, if there are no emotions attached to the various options in front of them, they will never reach a decision or conviction (Damasio 1994). This is critical for moral choice, because if anything morality involves strong convictions. These convictions don't—or rather can't—come about through a cool rationality: they require caring about others and powerful "gut feelings" about right and wrong.

Westermarck (1912 [1908], 1917 [1908]) discusses, one by one, a whole range of what philosophers before him, most notably David Hume (1985 [1739]), called the "moral sentiments." He classified the retributive emotions into those derived from resentment and anger, which seek revenge and punishment, and those that are more positive and prosocial. Whereas in his time few animal examples of the moral emotions were known—hence his reliance on Moroccan camel stories—we know now that there are many parallels in primate behavior. He also discusses "forgiveness," and how the turning of the other cheek is a universally appreciated gesture. Chimpanzees kiss and embrace after fights, and these so-called reconciliations serve to preserve peace within the community (de Waal and van Roosmalen 1979). A growing literature exists on conflict resolution in primates and other mammals (de Waal 1989b, 2000; Aureli and de Waal 2000; Aureli et al. 2002). Reconciliation may not be the same as forgiveness, but the two are obviously related.

Westermarck also sees protection of others against aggression as resulting from what he calls "sympathetic resentment," thus implying that this behavior rests on identification and empathy with the other. Protection against aggression is common in monkeys and apes and in many other animals, who stick up for their kin and friends. The primate literature offers a well-investigated picture of coalitions and alliances, which some consider the hallmark of primate social life and the main reason that primates have evolved such complex, cognitively demanding societies (e.g., Byrne and Whiten 1988; Harcourt and de Waal 1992; de Waal 1998 [1982]).

Similarly, the retributive kindly emotions ("desire to give pleasure in return for pleasure": Westermarck 1912 [1908]: 93) have an obvious parallel in what we now call reciprocal altruism, such as the tendency to repay in kind those from whom assistance has been received. Westermarck adds moral approval as a retributive kindly emotion, hence as a component of reciprocal altruism. These views antedate the discussions about "indirect reciprocity" in the modern literature on evolutionary ethics, which revolve around reputation building within the larger community (e.g., Alexander 1987). It is truly amazing to see how many issues brought up by contemporary authors are, couched in somewhat different terms, already present in the writings of this Swedish Finn of a century ago.

The most insightful part of Westermarck's work is perhaps where he tries to come to grips with what defines a moral emotion as moral. Here he shows that there is more to such emotions than raw gut feeling, as he explains that they "differ from kindred non-moral emotions by their disinterestedness, apparent impartiality, and flavour of generality" (Westermarck 1917 [1908]: 738–39). Emotions such as gratitude and resentment directly concern one's own interests—how one has been treated or how one wishes to be treated—hence they are too egocentric to be moral. Moral emotions ought to be disconnected from one's immediate situation: they deal with good and bad at a more abstract, disinterested level. It is only when we make general judgments of how *anyone* ought to be treated that we can begin to speak of moral approval and disapproval. It is in this specific area, famously symbolized by Smith's (1937 [1759]) "impartial spectator," that humans seem to go radically further than other primates.

Sections 4 and 5 discuss continuity between the two main pillars of human morality and primate behavior. Empathy and reciprocity have been described as the chief "prerequisites" (de Waal 1996) or "building blocks" of morality (Flack and de Waal 2000)—they are by no means sufficient to produce morality as we know it, yet they are indispensable. No human moral society could be imagined without reciprocal exchange and an emotional interest in others. This offers a concrete starting point to investigate the continuity that Darwin envisioned. The debate about Veneer Theory is fundamental to this investigation since some evolutionary biologists have sharply deviated from the idea of continuity by presenting morality as a sham so convoluted that only one species—ours—is capable of it. This view has no basis in fact, and as such stands in the way of a full understanding of how we became moral (table 1). My intention here is to set the record straight by reviewing actual empirical data.

ANIMAL EMPATHY

Evolution rarely throws out anything. Structures are transformed, modified, co-opted for other functions, or "tweaked" in another direction—descent with modification, as Darwin called it. Thus, the frontal fins of fish became the front limbs of land animals, which over time turned into hoofs, paws, wings, hands, and flippers. Occasionally, a structure loses all function and becomes superfluous, but this is a gradual process, often ending in rudimentary traits rather than disappearance. We find tiny vestiges of leg bones under the skin of whales and remnants of a pelvis in snakes.

This is why to the biologist, a Russian doll is such a satisfying plaything, especially if it has a historical dimension. I own a doll that shows Russian President Vladimir Putin on the outside, within whom we discover, in this order, Yeltsin, Gorbachev, Brezhnev, Kruschev, Stalin, and Lenin. Finding a little Lenin and Stalin within Putin will hardly surprise most

TABLE 1

Comparison of Veneer Theory and the View of Morality as an Outgrowth of the Social Instincts

	Veneer Theory	Evolution of Ethics
Origin	Huxleyan	Darwinian
Advocates	Richard Dawkins, George Williams, Robert Wright, etc.	Edward Westermarck, Edward Wilson, Jonathan Haidt, etc.
Type	Dualistic—pits humans against animals, and culture against nature. Morality is seen as a choice.	Unitary—postulates continuity between human morality and animal social tendencies. Moral tendencies are seen as evolved.
Proposed transition	From amoral animal to moral human	From social to moral animal
Theory	A position in search of a theory. It offers no explanation of why humans are "nicer than is good for their selfish genes," nor how such a feat might have been accomplished.	Theories of kin selection, reciprocal altruism, and their derivatives (e.g., fairness, reputation building, conflict resolution) suggest how a transition from social to moral animal might have come about.
Empirical evidence	None	a) Psychology—human morality has an emotional and intuitive foundation. b) Neuroscience—moral dilemmas activate emotionally involved brain areas. c) Primate behavior—our relatives show many of the tendencies incorporated into human morality.

political analysts. The same is true for biological traits: the old always remains present in the new.

This is relevant to the debate about the origin of empathy, since the psychologist tends to look at the world through different eyes than the biologist. Psychologists sometimes put our most advanced traits on a pedestal, ignoring or even denying simpler antecedents. They thus believe in saltatory change, at least in relation to our own species. This leads to unlikely origin stories, postulating discontinuities with respect to language, which is said to result from a unique "module" in the human brain (e.g., Pinker 1994), or with respect to human cognition, which is viewed as having cultural origins (e.g., Tomasello 1999). True, human capacities reach dizzying heights, such as when I understand that you understand that I understand, et cetera. But we are not born with such "reiterated empathy," as phenomenologists call it. Both developmentally and evolutionarily, advanced forms of empathy are preceded by and grow out of more elementary ones. In fact, things may be exactly the other way around. Instead of language and culture appearing with a Big Bang in our species and then transforming the way we relate to each other, Greenspan and Shanker (2004) propose that it is from early emotional connections and "proto conversations" between mother and child (cf. Trevarthen 1993) that language and culture sprang. Instead of empathy being an endpoint, it may have been the starting point.

Biologists prefer bottom-up over top-down accounts, even though there is definitely room for the latter. Once higher order processes have come into existence, they modify processes at the base. The central nervous system is a good example of top-down processing, as in the control the prefrontal cortex exerts over memory. The prefrontal cortex is not the seat of memory, but can "order" memory retrieval (Tomita et al. 1999). In the same way, culture and language shape expressions of empathy. The distinction between "being the origin of " and "shaping" is a fundamental one, though, and I will argue here that empathy is the original, pre-linguistic form of inter-individual linkage that only secondarily has come under the influence of language and culture.

Bottom-up accounts are the opposite of Big Bang theories. They assume continuity between past and present, child and adult, human and animal, even between humans and the most primitive mammals. We may assume that empathy first evolved in the context of parental care, which is obligatory in mammals (Eibl-Eibesfeldt 1974 [1971]; MacLean 1985). Signaling their state through smiling and crying, human infants urge their caregiver to pay attention and move into action (Bowlby 1958). The same applies to other primates. The survival value of these interactions is obvious. For example, a female chimpanzee lost a succession of infants despite intense positive interest because she was deaf and did not correct positional problems (such as sitting on the infant, or holding it upside-down) in response to its distress calls (de Waal 1998 [1982]).

For a human characteristic, such as empathy, that is so pervasive, develops so early in life (e.g., Hoffman 1975; Zahn-Waxler and Radke-Yarrow 1990), and shows such important neural and physiological correlates (e.g., Adolphs et al. 1994; Rimm-Kaufman & Kagan 1996; Decety and Chaminade 2003) as well as a genetic substrate (Plomin et al. 1993), it would be strange indeed if no evolutionary continuity existed with other mammals. The possibility of empathy and sympathy in other animals has been largely ignored, however. This is partly due to an excessive fear of anthropomorphism, which has stifled research into animal emotions (Panksepp 1998; de Waal 1999, appendix A), and partly to the one-sided portrayal by biologists of the natural world as a place of combat rather than social connectedness.

What Is Empathy?

Social animals need to coordinate action and movement, collectively respond to danger, communicate about food and water, and assist those in need. Responsiveness to the behavioral states of conspecifics ranges from a flock of birds taking off all at once because one among them is startled by a predator to a mother ape who returns to a whimpering youngster to help it from one tree to the next by draping her body as a bridge between the two. The first is a reflex-like transmission of fear that may not involve any understanding of what triggered the initial reaction, but that is undoubtedly adaptive. The bird that fails to take off at the same instant as the rest of the flock may be lunch. The selection pressure on paying attention to others must have been enormous. The mother-ape example is more discriminating, involving anxiety at hearing one's offspring whimper, assessment of the reason for its distress, and an attempt to ameliorate the situation.

There exists ample evidence of one primate coming to another's aid in a fight, putting an arm around a previous victim of attack, or other emotional responses to the distress of others (to be reviewed below). In fact, almost all communication among nonhuman primates is thought to be emotionally mediated. We are familiar with the prominent role of emotions in human facial expressions (Ekman 1982), but when it comes to monkeys and apes—which have a homologous array of expressions (van Hooff 1967)—emotions seem equally important.

When the emotional state of one individual induces a matching or closely related state in another, we speak of "emotional contagion" (Hatfield et al. 1993). Even if such contagion is undoubtedly a basic phenomenon, there is more to it than simply one individual being affected by the state of another: the two individuals often engage in direct interaction. Thus, a rejected youngster may throw a screaming tantrum at its mother's feet, or a preferred associate may approach a food possessor to beg by means of sympathy-inducing facial expressions, vocalizations, and hand gestures. In other words, emotional and motivational states often manifest themselves in behavior specifically directed at a partner. The emotional effect on the other is not a by-product, therefore, but actively sought.

With increasing differentiation between self and other, and an increasing appreciation of the precise circumstances underlying the emotional states of others, emotional contagion develops into empathy. Empathy encompasses—and could not possibly have arisen without—emotional contagion, but it goes beyond it in that it places filters between the other's and one's own state. In humans, it is around the age of two that we begin to add these cognitive layers (Eisenberg and Strayer 1987).

Two mechanisms related to empathy are *sympathy* and *personal distress*, which in their social consequences are each other's opposites. Sympathy is defined as "an affective response that consists of feelings of sorrow or concern for a distressed or needy other (rather than the same emotion as the other person). Sympathy is believed to involve an other-oriented, altruistic motivation" (Eisenberg 2000: 677). Personal distress, on the other hand, makes the affected party selfishly seek to alleviate its *own* distress, which is similar to what it has perceived in the object. Personal distress is therefore not concerned with the situation of the empathy-inducing other (Batson 1990). A striking primate example is given by de Waal (1996: 46): the screams of a severely punished or rejected infant rhesus monkey will often cause other infants to approach, embrace, mount, or even pile on top of the victim. Thus, the distress of one infant seems to spread to its peers, which then seek contact to soothe their own arousal. Inasmuch as personal distress lacks cognitive evaluation and behavioral complementarity, it does not reach beyond the level of emotional contagion.

That most modern textbooks on animal cognition (e.g., Shettleworth 1998) fail to index empathy or sympathy does not mean that these capacities are not an essential part of animal lives; it only means that they are being overlooked by a science traditionally focused on individual rather than inter-individual capacities. Tool use and numerical competence, for instance, are seen as hallmarks of intelligence, whereas appropriately dealing with others is not. It is obvious, however, that survival often depends on how animals fare within their group, both in a cooperative sense (e.g., concerted action, information transfer) and in a competitive sense (e.g., dominance strategies, deception). It is in the *social* domain, therefore, that one expects the highest cognitive achievements. Selection must have favored mechanisms to evaluate the emotional states of others and quickly respond to them. Empathy is precisely such a mechanism.

In human behavior, there exists a tight relation between

empathy and sympathy, and their expression in psychological altruism (e.g., Hornblow 1980; Hoffman 1982; Batson et al. 1987; Eisenberg and Strayer 1987; Wispé 1991). It is reasonable to assume that the altruistic and caring responses of other animals, especially mammals, rest on similar mechanisms. When Zahn-Waxler visited homes to find out how children respond to family members instructed to feign sadness (sobbing), pain (crying), or distress (choking), she discovered that children a little over one year of age already comfort others. Since expressions of sympathy emerge at an early age in virtually every member of our species, they are as natural as the first step. An unplanned sidebar to this study, however, was that household pets appeared as worried as the children by the "distress" of family members. They hovered over them or put their heads in their laps (Zahn-Waxler et al. 1984).

Rooted in attachment and what Harlow termed the "affectional system" (Harlow and Harlow 1965), responses to the emotions of others are commonplace in social animals. Thus, behavioral and physiological data suggest emotional contagion in a variety of species (reviewed in Preston and de Waal 2002b, and de Waal 2003). An interesting literature that appeared in the 1950s and '60s by experimental psychologists placed the words "empathy" and "sympathy" between quotation marks. In those days, talk of animal emotions was taboo. In a paper provocatively entitled "Emotional Reactions of Rats to the Pain of Others," Church (1959) established that rats that had learned to press a lever to obtain food would stop doing so if their response was paired with the delivery of an electric shock to a visible neighboring rat. Even though this inhibition habituated rapidly, it suggested something aversive about the pain reactions of others. Perhaps such reactions arouse negative emotions in rats that witness them.

Monkeys show a stronger inhibition than rats. The most compelling evidence for the strength of empathy in monkeys came from Wechkin et al. (1964) and Masserman et al. (1964). They found that rhesus monkeys refuse to pull a chain that delivers food to themselves if doing so shocks a companion. One monkey stopped pulling for five days, and another one for twelve days after witnessing shock delivery to a companion. These monkeys were literally starving themselves to avoid inflicting pain upon another. Such sacrifice relates to the tight social system and emotional linkage among these macaques, as supported by the finding that the inhibition to hurt another was more pronounced between familiar than unfamiliar individuals (Masserman et al. 1964).

Although these early studies suggest that, by behaving in certain ways, animals try to alleviate or prevent distress in others, it remains unclear if spontaneous responses to distressed conspecifics are explained by (a) aversion to distress signals of others, (b) personal distress generated through emotional contagion, or (c) true helping motivations. Work on nonhuman primates has furnished further information. Some of this evidence is qualitative, but quantitative data on empathic reactions exists as well.

Anecdotes of "Changing Places in Fancy"

Striking depictions of primate empathy and altruism can be found in Yerkes (1925), Ladygina-Kohts (2002 [1935]), Goodall (1990), and de Waal (1998 [1982], 1996, 1997a). Primate empathy is such a rich area that O'Connell (1995) was able to conduct a content analysis of thousands of qualitative reports. She concluded that responses to the distress of another seem considerably more complex in apes than monkeys. To give just one example of the strength of the ape's empathic response, Ladygina-Kohts wrote about her young chimpanzee, Joni, that the best way to get him off the roof of her house (much better than any reward or threat of punishment) was by arousing his sympathy:

> If I pretend to be crying, close my eyes and weep, Joni immediately stops his plays or any other activities, quickly runs over to me, all excited and shagged, from the most remote places in the house, such as the roof or the ceiling of his cage, from where I could not drive him down despite my persistent calls and entreaties. He hastily runs around me, as if looking for the offender; looking at my face, he tenderly takes my chin in his palm, lightly touches my face with his finger, as though trying to understand what is happening, and turns around, clenching his toes into firm fists. (Ladygina-Kohts, 2002 [1935]: 121)

De Waal (1996, 1997a) has suggested that apart from emotional connectedness, apes have an appreciation of the other's situation and a degree of perspective-taking (appendix B). So, the main difference between monkeys and apes is not in empathy *per se*, but in the cognitive overlays, which allow apes to adopt the other's viewpoint. One striking report in this regard concerns a bonobo female empathizing with a bird at Twycross Zoo, in England:

> One day, Kuni captured a starling. Out of fear that she might molest the stunned bird, which appeared undamaged, the

keeper urged the ape to let it go. . . . Kuni picked up the starling with one hand and climbed to the highest point of the highest tree where she wrapped her legs around the trunk so that she had both hands free to hold the bird. She then carefully unfolded its wings and spread them wide open, one wing in each hand, before throwing the bird as hard she could towards the barrier of the enclosure. Unfortunately, it fell short and landed onto the bank of the moat where Kuni guarded it for a long time against a curious juvenile. (de Waal, 1997a, p. 156)

What Kuni did would obviously have been inappropriate towards a member of her own species. Having seen birds in flight many times, she seemed to have a notion of what would be good for a bird, thus offering us an anthropoid version of the empathic capacity so enduringly described by Adam Smith (1937 [1759]: 10) as "changing places in fancy with the sufferer." Perhaps the most striking example of this capacity is a chimpanzee who, as in the original Theory of Mind (ToM) experiments of Premack and Woodruff (1978), seemed to understand the intentions of another and provided specific assistance:

During one winter at the Arnhem Zoo, after cleaning the hall and before releasing the chimps, the keepers hosed out all rubber tires in the enclosure and hung them one by one on a horizontal log extending from the climbing frame. One day, Krom was interested in a tire in which water had stayed behind. Unfortunately, this particular tire was at the end of the row, with six or more heavy tires hanging in front of it. Krom pulled and pulled at the one she wanted but couldn't remove it from the log. She pushed the tire backward, but there it hit the climbing frame and couldn't be removed either. Krom worked in vain on this problem for over ten minutes, ignored by everyone, except Jakie, a seven-year-old Krom had taken care of as a juvenile.

Immediately after Krom gave up and walked away, Jakie approached the scene. Without hesitation he pushed the tires one by one off the log, beginning with the front one, followed by the second in the row, and so on, as any sensible chimp would. When he reached the last tire, he carefully removed it so that no water was lost, carrying it straight to his aunt, placing it upright in front of her. Krom accepted his present without any special acknowledgment, and was already scooping up water with her hand when Jakie left. (Adapted from de Waal 1996)

That Jakie assisted his aunt is not so unusual. What is special is that he correctly guessed what Krom was after. He grasped his auntie's goals. Such so-called "targeted helping" is typical of apes, but rare or absent in most other animals. It is defined as altruistic behavior tailored to the specific needs of the other even in novel situations, such as the highly publicized case of Binti Jua, a female gorilla who rescued a human child at the Brookfield Zoo in Chicago (de Waal, 1996, 1999). A recent experiment demonstrated targeted helping in young chimpanzees (Warneken and Tomasello 2006).

It is important to stress the incredible strength of the ape's helping response, which makes these animals take great risks on behalf of others. Whereas in a recent debate about the origins of morality, Kagan (2000) considered it obvious that a chimpanzee would never jump into a cold lake to save another, it may help to quote Goodall (1990: 213) on this issue:

In some zoos, chimpanzees are kept on man-made islands, surrounded by water-filed moats. . . . Chimpanzees cannot swim and, unless they are rescued, will drown if they fall into deep water. Despite this, individuals have sometimes made heroic efforts to save companions from drowning— and were sometimes successful. One adult male lost his life as he tried to rescue a small infant whose incompetent mother had allowed it to fall into the water.

The only other animals with a similar array of helping responses are dolphins and elephants. This evidence, too, is largely descriptive (dolphins: Caldwell and Caldwell 1966; Connor and Norris 1982; elephants: Moss 1988; Payne 1998), yet here again it is hard to accept as coincidental that scientists who have watched these animals for any length of time have numerous such stories, whereas scientists who have watched other animals have few, if any.

Consolation Behavior

This difference between monkey and ape empathy has been confirmed by systematic studies of a behavior known as "consolation," first documented by de Waal and van Roosmalen (1979). Consolation is defined as reassurance by an uninvolved bystander to one of the combatants in a preceding aggressive incident. For example, a third party goes over to the loser of a fight and gently puts an arm around his or her shoulders (figure 2). Consolation is not to be confused with reconciliation between former opponents, which seems mostly motivated by self-interest, such as the imperative to restore a disturbed social relationship (de Waal 2000). The advantage of consolation for the actor remains wholly

Figure 2 A typical instance of consolation in chimpanzees in which a juvenile puts an arm around a screaming adult male who has just been defeated in a fight with his rival. Photograph by the author.

unclear. The actor could probably walk away from the scene without any negative consequences.

Information on chimpanzee consolation is well quantified. De Waal and van Roosmalen (1979) based their conclusions on an analysis of hundreds of postconflict observations, and a replication by de Waal and Aureli (1996) included an even larger sample in which the authors sought to test two relatively simple predictions. If third-party contacts indeed serve to alleviate the distress of conflict participants, these contacts should be directed more at recipients

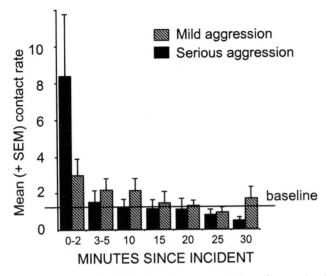

Figure 3 The rate at which third parties contact victims of aggression in chimpanzees, comparing recipients of serious and mild aggression. Especially in the first few minutes after the incident, recipients of serious aggression receive more contacts than baseline. After de Waal and Aureli (1996).

of aggression than at aggressors, and more at recipients of intense rather than mild aggression. Comparing third-party contact rates with baseline levels, the investigators found support for both predictions (figure 3).

Consolation has thus far been demonstrated in great apes only. When de Waal and Aureli (1996) set out to apply exactly the same observation methodology as used on chimpanzees to detect consolation in macaques, they failed to find any (reviewed by Watts et al. 2000). This came as a surprise, because reconciliation studies, which employ essentially the same data collection method, have shown reconciliation in species after species. Why, then, would consolation be restricted to apes?

Possibly, one cannot achieve cognitive empathy without a high degree of self-awareness. Targeted help in response to specific, sometimes novel, situations may require a distinction between self and other that allows the other's situation to be divorced from one's own while maintaining the emotional link that motivates behavior. In other words, in order to understand that the source of vicarious arousal is not oneself but the other and to understand the causes of the other's state, one needs a clear distinction between self and other. Based on these assumptions, Gallup (1982) was the first to speculate about a connection between cognitive empathy and mirror self-recognition (MSR). This view is supported both developmentally, by a correlation between the emergence of MSR in young children and their helping tendencies (Bischof-Köhler 1988; Zahn-Waxler et al. 1992), and phylogenetically, by the presence of complex helping and consolation in hominoids (i.e., humans and apes) but not monkeys. Hominoids are also the only primates with MSR.

I have argued before that, apart from consolation behavior, targeted helping reflects cognitive empathy. Targeted helping is defined as altruistic behavior tailored to the specific needs of the other in novel situations, such as the previously described reaction of Kuni to the bird or Binti Jua's rescue of a boy. These responses require an understanding of the specific predicament of the individual needing help. Given the evidence for targeted helping by dolphins (see above), the recent discovery of MSR in these mammals (Reiss and Marino 2001) supports the proposed connection between increased self-awareness, on the one hand, and cognitive empathy, on the other.

Russian Doll Model

The literature includes accounts of empathy as a cognitive affair, even to the point that apes, let alone other animals, probably lack it (Povinelli 1998; Hauser 2000). This view equates empathy with mental state attribution and ToM. The opposite position has recently been defended in relation to autistic children, however. Contra earlier assumptions that autism reflects a ToM deficit (Baron-Cohen 2000), autism is noticeable well before the age of 4 years at which ToM typically emerges. Williams et al. (2001) argue that the main deficit of autism concerns the socio-affective level, which in turn negatively impacts sophisticated downstream forms of interpersonal perception, such as ToM. Thus, ToM is seen as a derived trait, and the authors urge more attention to its antecedents (a position now also embraced by Baron-Cohen 2003, 2004).

Preston and de Waal (2002a) propose that at the core of the empathic capacity is a relatively simple mechanism that provides an observer (the "subject") with access to the emotional state of another (the "object") through the subject's own neural and bodily representations. When the subject attends to the object's state, the subject's neural representations of similar states are automatically activated. The closer and more similar subject and object are, the easier it will be for the subject's perception to activate motor and autonomic responses that match the object's (e.g., changes in heart rate, skin conductance, facial expression, body posture). This activation allows the subject to get "under the skin" of the object, sharing its feelings and needs, which embodiment in turn fosters sympathy, compassion, and helping. Preston and de Waal's (2002a) Perception-Action Mechanism (PAM) fits Damasio's (1994) somatic marker hypothesis of emotions as well as recent evidence for a link at the cellular level between perception and action (e.g., "mirror neurons," di Pelligrino et al. 1992).

The idea that perception and action share representations is anything but new: it goes as far back as the first treatise on *Einfühlung*, the German concept translated into English as "empathy" (Wispé 1991). When Lipps (1903) spoke of *Einfühlung*, which literally means "feeling into," he speculated about *innere Nachahmung* (inner mimicry) of another's feelings along the same lines as proposed by the PAM. Accordingly, empathy is a routine involuntary process, as demonstrated by electromyographic studies of invisible muscle contractions in people's faces in response to pictures of human facial expressions. These reactions are fully automated and occur even when people are unaware of what they saw (Dimberg et al. 2000). Accounts of empathy as a higher cognitive process neglect these gut-level reactions, which are far too rapid to be under conscious control.

Perception-action mechanisms are well known for motor perception (Prinz and Hommel 2002), causing researchers to assume similar processes to underlie emotion perception (Gallese 2001; Wolpert et al. 2001). Data suggest that both observing and experiencing emotions involves shared physiological substrates: *seeing* another's disgust or pain is very much like *being* disgusted or in pain (Adolphs et al. 1997, 2000; Wicker et al. 2003). Also, affective communication creates similar physiological states in subject and object (Dimberg 1982, 1990; Levenson and Reuf 1992). In short, human physiological and neural activity does not take place on an island, but is intimately connected with and affected by

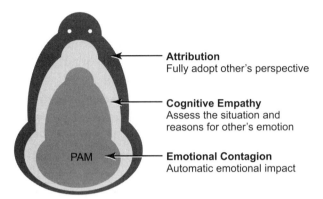

Figure 4 According to the Russian Doll Model, empathy covers all processes leading to related emotional states in subject and object. At its core is a simple, automatic Perception-Action Mechanism (PAM), which results in immediate, often unconscious state matching between individuals. Higher levels of empathy that build on this hardwired basis include cognitive empathy (i.e., understanding the reasons for the other's emotions) and mental state attribution (i.e., fully adopting the other's perspective). The Russian Doll Model proposes that outer layers require inner ones. After de Waal (2003).

fellow human beings. Recent investigations of the neural basis of empathy lend strong support to the PAM (Carr et al. 2003; Singer et al. 2004; de Gelder et al. 2004).

How simple forms of empathy relate to more complex ones has been depicted as a Russian doll by de Waal (2003). Accordingly, empathy covers all forms of one individual's emotional state affecting another's, with basic mechanisms at its core and more advanced mechanisms and cognitive abilities as its outer layers (figure 4). Autism may be reflected in deficient outer layers of the Russian doll, but such deficiencies invariably go back to deficient inner layers.

This is not to say that higher cognitive levels of empathy

are irrelevant, but they are built on top of this firm, hardwired basis without which we would be at a loss about what moves others. Surely, not all empathy is reducible to emotional contagion, but it never gets around it. At the core of the Russian doll, we find a PAM-induced emotional state that corresponds with the object's state. In a second layer, cognitive empathy implies appraisal of another's predicament or situation (cf. de Waal 1996). The subject not only responds to the signals emitted by the object, but seeks to understand the reasons for these signals, looking for clues in the other's behavior and situation. Cognitive empathy makes it possible to furnish targeted help that takes the specific needs of the other into account (figure 5). These responses go well beyond emotional contagion, yet they would be hard to explain without the motivation provided by the emotional component. Without it, we would be as disconnected as Mr. Spock in *Star Trek*, constantly wondering why others feel what they say they feel.

Whereas monkeys (and many other social mammals) clearly seem to possess emotional contagion and a limited degree of targeted helping, the latter phenomenon is not nearly as robust as in the great apes. For example, at Jigokudani Monkey Park, in Japan, first-time mother macaques are kept out of the hot springs by park wardens because of the experience that these females will accidentally drown their infants. They fail to pay attention to them when submerging themselves in the ponds. This is something monkey mothers apparently have to learn with time, showing that they do not automatically take their offspring's perspective. De Waal (1996) ascribed their behavioral change to "learned adjustment," setting it apart from cognitive empathy, which is more typical of apes and humans. Ape mothers respond immediately and appropriately to the specific needs of their

Figure 5 Cognitive empathy (i.e., empathy combined with appraisal of the other's situation) allows for aid tailored to the other's needs. In this case, a mother chimpanzee reaches out to help her son out of a tree after he has screamed and begged (see hand gesture). Targeted helping may require a distinction between self and other, an ability also thought to underlie mirror self-recognition, as found in humans, apes, and dolphins. Photograph by the author.

offspring. They are, for example, very careful to keep them away from water, rushing over to pull them away as soon as they get too close.

In conclusion, empathy is not an all-or-nothing phenomenon: it covers a wide range of emotional linkage patterns, from the very simple and automatic to the highly sophisticated. It seems logical to first try to understand the basic forms of empathy, which are widespread indeed, before addressing the variations that cognitive evolution has constructed on top of this foundation.

RECIPROCITY AND FAIRNESS

Chimpanzees and capuchin monkeys—the two species I work with most—are special, as they are among the very few primates that share food outside the mother-offspring context (Feistner and McGrew 1989). The capuchin is a small primate, easy to work with, as opposed to the chimpanzee, which is many times stronger than we are. Members of both species are interested in each other's food and will share food on occasion—sometimes even hand over a piece to another. Most sharing, however, is passive, where one individual will reach for food owned by another, who will let go. But even passive sharing is special when compared to most animals, for which a similar situation would result in a fight or assertion by the dominant, without any sharing at all.

Chimpanzee Gratitude

We studied sequences involving food sharing to see how a beneficial act by individual A toward B would affect B's behavior toward A. The prediction was that B would show beneficial behavior toward A in return. The problem with food sharing is, however, that after a group-wide feeding session as used in our experiments, the motivation to share changes (the animals are more sated). Hence, food sharing cannot be the only variable measured. A second social service unaffected by food consumption was included. For this, grooming between individuals prior to food sharing was used. The frequency and duration of hundreds of spontaneous grooming bouts among our chimpanzees were measured in the morning. Within half an hour after the end of these observations, starting around noon, the apes were given two tightly bound bundles of leaves and branches. Nearly 7,000 interactions over food were carefully recorded by observers and entered into a computer according to strict definitions described by de Waal (1989a). The resulting database on spontaneous services exceeds that for any other nonhuman primate.

It was found that adults were more likely to share food with individuals who had groomed them earlier. In other words, if A had groomed B in the morning, B was more likely than usual to share food with A later in the day. This result, however, could be explained in two ways. The first is the "good mood" hypothesis according to which individuals who have received grooming are in a benevolent mood, leading them to share indiscriminately with all individuals. The second explanation is the direct-exchange hypothesis, in which the individual who has been groomed responds by sharing food specifically with the groomer. The data indicated that the sharing increase was specific to the previous groomer. In other words, chimpanzees appeared to remember others who had just performed a service (grooming) and respond to those individuals by sharing more with them. Also, aggressive protests by food possessors to approaching individuals were directed more at those who had not groomed them than at previous grooming partners. This is compelling evidence for partner-specific reciprocal exchange (de Waal 1997b).

Of all existing examples of reciprocal altruism in nonhuman animals, the exchange of food for grooming in chimpanzees appears to be the most cognitively advanced. Our
data strongly suggest a memory-based mechanism. A significant time delay existed between favors given and received
(from half an hour to two hours); hence the favor was acted
upon well after the previous interaction. Apart from memory of past events, we need to postulate that the memory of
a received service, such as grooming, triggered a positive attitude toward the individual who offered this service, a psychological mechanism known in humans as "gratitude."
Gratitude within the context of reciprocal exchange was pre-
dicted by Trivers (1971), and has been discussed by Bonnie
and de Waal (2004). It is classified by Westermarck (1912
[1908]) as one of the "retributive kindly emotions" deemed
essential for human morality.

Monkey Fairness

During the evolution of cooperation it may have become
critical for actors to compare their own efforts and payoffs
with those of others. Negative reactions may ensue in case of
violated expectations. A recent theory proposes that aversion to inequity can explain human cooperation within the
bounds of the rational choice model (Fehr and Schmidt
1999). Similarly, cooperative nonhuman species seem guided
by a set of expectations about the outcome of cooperation
and access to resources. De Waal (1996: 95) proposed a *sense
of social regularity*, defined as "A set of expectations about
the way in which oneself (or others) should be treated and
how resources should be divided. Whenever reality deviates

from these expectations to one's (or the other's) disadvantage, a negative reaction ensues, most commonly protest by subordinate individuals and punishment by dominant individuals."

The sense of how others should or should not behave is essentially egocentric, although the interests of individuals close to the actor, especially kin, may be taken into account (hence the parenthetical inclusion of others). Note that the expectations have not been specified: they tend to be species-typical. For example, a rhesus monkey expects no share of a dominant's food, as it lives in a despotically hierarchical society, but a chimpanzee definitely does, hence the begging, whining, and temper tantrums if no share is forthcoming. I consider expectations the most important unstudied topic in animal behavior, which is all the more lamentable as it is the one issue that will bring animal behavior closest to the "ought" of behavior that we recognize so clearly in the moral domain.

To explore the expectations held by capuchin monkeys, we made use of their ability to judge and respond to value. We knew from previous studies that capuchins easily learn to assign value to tokens. Furthermore they can use these assigned values to complete a simple barter. This allowed a test to elucidate inequity aversion by measuring the reactions of subjects to a partner receiving a superior reward for the same tokens.

We paired each monkey with a group mate and watched their reactions when their partners got a better reward for doing the same bartering task. This consisted of an exchange in which the experimenter gave the subject a token that could immediately be handed back for a reward (figure 6). Each session consisted of twenty-five exchanges by each individual,

Figure 6 A capuchin monkey in the test chamber returns a token to the experimenter with her right hand while steadying the human hand with her left hand. Her partner looks on. Drawing by Gwen Bragg and Frans de Waal after a video still.

and the subject always saw the partner's exchange immediately before its own. Food rewards varied from lower value rewards (e.g., a cucumber piece), which they are usually happy to work for, to higher value rewards (e.g., a grape), which were preferred by all individuals tested. All subjects were subjected to (a) an Equity Test (ET), in which subject and partner did the same work for the same low-value food, (b) an Inequity Test (IT), in which the partner received a superior reward (grape) for the same effort, (c) an Effort Control Test (EC), designed to elucidate the role of effort, in

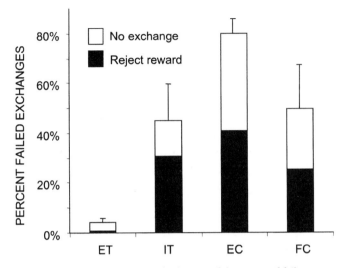

Figure 7 Mean percentage ± standard error of the mean of failures to exchange for females across the four test types. Black bars represent the proportion of nonexchanges due to refusals to accept the reward; white bars represent nonexchanges due to refusals to return the token. ET = Equity Test, IT = Inequity Test, EC = Effort Control, FC = Food Control. The Y-axis shows the percentage of nonexchanges.

which the partner received the higher value grape for free, and (d) a Food Control Test (FC), designed to elucidate the effect of the presence of the reward on subject behavior, in which grapes were visible but not given to another capuchin.

Individuals who received lower value rewards showed both passive negative reactions (e.g., refusing to exchange the token, ignoring the reward) and active negative reactions (e.g., throwing out the token or the reward). Compared to tests in which both received identical rewards, the capuchins were far less willing to complete the exchange or accept the reward if their partner received a better deal (figure 7; Brosnan and de Waal 2003). Capuchins refused to participate even more frequently if their partner did not have to work (exchange) to get the better reward but was handed it for "free." Of course, there is always the possibility that subjects were just reacting to the presence of the higher value food and that what the partner received (free or not) did not affect their reaction. However, in the Food Control Test, in which the higher value reward was visible but not given to another monkey, the reaction to the presence of this high-value food decreased significantly over the course of testing, which is a change in the opposite direction from that seen when the high-value reward went to an actual partner. Clearly our subjects discriminate between higher value food being consumed by a conspecific and such food being merely visible, intensifying their rejections only to the former (Brosnan and de Waal 2003).

Capuchin monkeys thus seem to measure reward in relative terms, comparing their own rewards with those available and their own efforts with those of others. Although our data cannot elucidate the precise motivations underlying these responses, one possibility is that monkeys, like humans, are guided by social emotions. In humans, these emotions, known as "passions" by economists, guide an individual's reactions to the efforts, gains, losses, and attitudes of others (Hirschleifer 1987; Frank 1988; Sanfey et al. 2003). As opposed to primates marked by despotic hierarchies (such as rhesus monkeys), tolerant species with well-developed food sharing and cooperation (such as capuchin monkeys) may hold emotionally charged expectations about reward distribution and social exchange that lead them to dislike inequity.

Before we speak of "fairness" in this context it is good to point out a difference between this and human fairness, though. A full-blown sense of fairness would entail that the "rich" monkey share with the "poor" one, as she should feel she is getting excessive compensation. Such behavior would betray interest in a higher principle of fairness, one that Westermarck (1917 [1908]) called "disinterested," hence a truly moral notion. This is not the sort of reaction our monkeys showed, though: their sense of fairness, if we call it that, was rather egocentric. They showed an expectation about how they themselves should be treated, not about how everybody around them should be treated. At the same time, it cannot be denied that the full-blown sense of fairness must have started someplace and that the self is the logical place to look for its origin. Once the egocentric form exists, it can be expanded to include others.

MENCIUS AND THE PRIMACY OF AFFECT

There is never much new under the sun. Westermarck's emphasis on the retributive emotions, whether friendly or vengeful, reminds one of the reply of Confucius to the question whether there is any single word that may serve as prescription for all of one's life. Confucius proposed "reciprocity" as such a word. Reciprocity is of course also at the heart of the Golden Rule, which remains unsurpassed as a summary of human morality. To know that some of the psychology behind this rule may exist in other species, along with the required empathy, bolsters the idea that morality, rather than a recent invention, is part of human nature.

A follower of Confucius, Mencius, wrote extensively about human goodness during his life, from 372 to 289 BC. Mencius lost his father when he was three, and his mother made sure he received the best possible education. The mother is at least as well known as her son: to the Chinese, she still serves as a maternal model for her absolute devotion. Called the "second sage" because of his immense influence, second only to Confucius, Mencius had a revolutionary, subversive bent in that he stressed the obligation of rulers to provide for the common people. Recorded on bamboo clappers and handed down to his descendants and their students, his writings show that the debate about whether we are naturally moral or not is ancient indeed. In one exchange, Mencius (n.d. [372–289 BC]: 270–71) reacts against Kaou Tsze's views, which are reminiscent of Huxley's gardener and garden metaphor:

> "Man's nature is like the *ke* willow, and righteousness is like a cup or a bowl. The fashioning of benevolence and righteousness out of man's nature is like the making of cups and bowls from the *ke* willow."

Mencius replied:

> "Can you, leaving untouched the nature of the willow, make with it cups and bowls? You must do violence and injury to the willow, before you can make cups and bowls with it. If you must do violence and injury to the willow, before you can make cups and bowls with it, on your principles you must in the same way do violence and injury to humanity in order to fashion from it benevolence and righteousness! Your words alas! would certainly lead all men on to reckon benevolence and righteousness to be calamities."

Mencius believed that humans tend toward the good as naturally as water flows downhill. This is also evident from the following remark, in which he seeks to exclude the possibility of the Freudian double agenda between presented and felt motives on the grounds that the immediacy of the moral emotions, such as sympathy, leaves no room for cognitive contortions:

> When I say that all men have a mind which cannot bear to see the suffering of others, my meaning may be illustrated thus: even nowadays, if men suddenly see a child about to fall into a well, they will without exception experience a feeling of alarm and distress. They will feel so, not as a ground on which they may gain the favor of the child's parents, nor as a ground on which they may seek the praise of their neighbors and friends, nor from a dislike to the reputation of having been unmoved by such a thing. From this case we may perceive that the feeling of commiseration is essential to man. (Mencius n.d. [372–289 BC]: 78)

This example from Mencius reminds us of Westermarck's epigraph ("Can we help sympathizing with our friends?") and the quotation from Smith ("How selfish soever man may be supposed . . ."). The central idea underlying all three statements is that distress at the sight of another's pain is an impulse over which we exert little or no control: it grabs us instantaneously, like a reflex, without time to weigh the pros and cons. All three statements hint at an involuntary process such as PAM. Remarkably, the possible alternative motives brought up by Mencius also feature in the modern literature, usually under the heading of reputation building. The big difference is, of course, that Mencius rejected these explanations as too contrived, given the immediacy and force of the sympathetic impulse. Manipulation of public opinion is entirely possible at other times, he said, but not at the very instant that a child falls into a well.

I could not agree more. Evolution has produced species

that follow genuinely cooperative impulses. I don't know if people are, deep down, good or evil, but to believe that each and every move is selfishly calculated, while being hidden from others (and often from ourselves), seems to grossly overestimate human intellectual powers, let alone those of other animals. Apart from the already discussed animal examples of consolation of distressed individuals and protection against aggression, there exists a rich literature on human empathy and sympathy that, generally, agrees with the assessment of Mencius that impulses in this regard come first and rationalizations later (e.g., Batson 1990; Wispé 1991).

COMMUNITY CONCERN

In this essay, I have drawn a stark contrast between two schools of thought on human goodness. One school sees people as essentially evil and selfish, and hence morality as a mere cultural overlay. This school, personified by T. H. Huxley, is still very much with us even though I have noticed that no one (not even those explicitly endorsing this position) likes to be called a "veneer theorist." This may be due to wording, or because once the assumptions behind Veneer Theory are laid bare, it becomes obvious that—unless one is willing to go the purely rationalist route of modern Hobbesians, such as Gauthier (1986)—the theory lacks any sort of explanation of how we moved from being amoral animals to moral beings. The theory is at odds with the evidence for emotional processing as driving force behind moral judgment. If human morality could truly be reduced to calculations and reasoning, we would come close to being psychopaths, who indeed do not mean to be kind when they act kindly. Most of us hope to be slightly better than that, hence the possible aversion to my black-and-white contrast between Veneer Theory and the alternative school, which seeks to ground morality in human nature.

This school sees morality arise naturally in our species and believes that there are sound evolutionary reasons for the capacities involved. Nevertheless, the theoretical framework to explain the transition from social animal to moral human consists only of bits and pieces. Its foundations are the theories of kin selection and reciprocal altruism, but it is obvious that other elements will need to be added. If one reads up on reputation building, fairness principles, empathy, and conflict resolution (in disparate literatures that cannot be reviewed here), there seems a promising movement toward a more integrated theory of how morality may have come about (see Katz 2000).

It should further be noted that the evolutionary pressures responsible for our moral tendencies may not all have been nice and positive. After all, morality is very much an ingroup phenomenon. Universally, humans treat outsiders far worse than members of their own community: in fact, moral rules hardly seem to apply to the outside. True, in modern times there is a movement to expand the circle of morality, and to include even enemy combatants—e.g., the Geneva Convention, adopted in 1949—but we all know how fragile an effort this is. Morality likely evolved as a within-group phenomenon in conjunction with other typical within-group capacities, such as conflict resolution, cooperation, and sharing.

The first loyalty of every individual is not to the group, however, but to itself and its kin. With increasing social integration and reliance on cooperation, shared interests must have risen to the surface so that the community as a whole became an issue. The biggest step in the evolution of human morality was the move from interpersonal relations to a focus on the greater good. In apes, we can see the beginnings of this when they smooth relations between others. Females may bring males together after a fight between them, thus brokering a reconciliation, and high-ranking males often stop fights among others in an evenhanded manner, thus promoting peace in the group. I see such behavior as a reflection of *community concern* (de Waal 1996), which in turn reflects the stake each group member has in a cooperative atmosphere. Most individuals have much to lose if the community were to fall apart, hence the interest in its integrity and harmony. Discussing similar issues, Boehm (1999) added the role of social pressure, at least in humans: the entire community works at rewarding group-promoting behavior and punishing group-undermining behavior.

Obviously, the most potent force to bring out a sense of community is enmity toward outsiders. It forces unity among elements that are normally at odds. This may not be visible at the zoo, but it is definitely a factor for chimpanzees in the wild, which show lethal intercommunity violence (Wrangham and Peterson 1996). In our own species, nothing is more obvious than that we band together against adversaries. In the course of human evolution, out-group hostility enhanced in-group solidarity to the point that morality emerged. Instead of merely ameliorating relations around us, as apes do, we have explicit teachings about the value of the community and the precedence it takes, or ought to take, over individual interests. Humans go much further in all of this than the apes (Alexander 1987), which is why we have moral systems and apes do not.

And so, the profound irony is that our noblest achievement—morality—has evolutionary ties to our basest behavior—warfare. The sense of community required by the former was provided by the latter. When we passed the tipping point between conflicting individual interests and shared interests, we ratcheted up the social pressure to make sure everyone contributed to the common good.

If we accept this view of an evolved morality, of morality as a logical outgrowth of cooperative tendencies, we are not going against our own nature by developing a caring, moral attitude, any more than civil society is an out-of-control garden subdued by a sweating gardener, as Huxley (1989 [1894]) thought. Moral attitudes have been with us from the start, and the gardener rather is, as Dewey aptly put it, an organic grower. The successful gardener creates conditions and introduces plant species that may not be normal for this particular plot of land "but fall within the wont and use of nature as a whole" (Dewey 1993 [1898]: 109–10). In other words, we are not hypocritically fooling everyone when we act morally: we are making decisions that flow from social instincts older than our species, even though we add to these the uniquely human complexity of a disinterested concern for others and for society as a whole.

Following Hume (1985 [1739]), who saw reason as the slave of the passions, Haidt (2001) has called for a thorough reevaluation of the role played by rationality in moral judgment, arguing that most human justification seems to occur *post hoc*, that is, after moral judgments have been reached on the basis of quick, automated intuitions. Whereas Veneer Theory, with its emphasis on human uniqueness, would predict that moral problem solving is assigned to evolutionarily recent additions to our brain, such as the prefrontal cortex, neuroimaging shows that moral judgment in fact involves a wide variety of brain areas, some extremely ancient (Greene and Haidt 2002). In short, neuroscience seems to be lending support to human morality as evolutionarily anchored in mammalian sociality.

We celebrate rationality, but when push comes to shove we assign it little weight (Macintyre 1999). This is especially true in the moral domain. Imagine that an extraterrestrial consultant instructs us to kill people as soon as they come down with influenza. In doing so, we are told, we would kill far fewer people than would die if the epidemic were allowed to run its course. By nipping the flu in the bud, we would save lives. Logical as this may sound, I doubt that many of us would opt for this plan. This is because human morality is firmly anchored in the social emotions, with empathy at its core. Emotions are our compass. We have strong inhibitions against killing members of our own community, and our moral decisions reflect these feelings. For the same reasons, people object to moral solutions that involve hands-on harm to another (Greene and Haidt 2002). This may be because hands-on violence has been subject to natural selection, whereas utilitarian deliberations have not.

Additional support for an intuitionist approach to morality comes from child research. Developmental psychologists used to believe that the child learns its first moral distinctions through fear of punishment and a desire for praise. Similar to veneer theorists, they conceived morality as coming from the outside, imposed by adults upon a passive, naturally selfish child. Children were thought to adopt parental values to construct a superego: the moral agency of the self. Left to their own devices, children would never arrive at anything close to morality. We know now, however, that at an early age children understand the difference between moral principles ("do not steal") and cultural conventions ("no pajamas at school"). They apparently appreciate that the breaking of certain rules distresses and harms others, whereas the breaking of other rules merely violates expectations about what is appropriate. Their attitudes don't seem based purely on reward and punishment. Whereas many pediatric handbooks still depict young children as self-centered monsters, it has become clear that by one year of age they spontaneously comfort others in distress (Zahn-Waxler et al. 1992) and that soon thereafter they begin to develop a moral perspective through interactions with other members of their species (Killen and Nucci 1995).

Instead of our doing "violence to the willow," as Mencius called it, to create the cups and bowls of an artificial morality, we rely on natural growth in which simple emotions, like those encountered in young children and social animals, develop into the more refined, other-including sentiments that we recognize as underlying morality. My own argument here obviously revolves around the continuity between human social instincts and those of our closest relatives, the monkeys and apes, but I feel that we are standing at the threshold of a much larger shift in theorizing that will end up positioning morality firmly within the emotional core of human nature. Humean thinking is making a major comeback.

Why did evolutionary biology stray from this path during the final quarter of the twentieth century? Why was morality considered unnatural, why were altruists depicted as hypocrites, and why were emotions left out of the debate? Why the calls to go against our own nature and to distrust a "Darwinian world"? The answer lies in what I have called the *Beethoven error*. In the same way that Ludwig van Beethoven is said to have produced his beautiful, intricate compositions in one of the most disorderly and dirty apartments of Vienna, there is not much of a connection between the process of natural selection and its many products. The Beethoven error is to think that, since natural selection is a cruel, pitiless process of elimination, it can only have produced cruel and pitiless creatures (de Waal 2005).

But nature's pressure cooker does not work that way. It favors organisms that survive and reproduce, pure and simple. How they accomplish this is left open. Any organism that can do better by becoming either more or less aggressive than the rest, more or less cooperative, or more or less caring, will spread its genes.

The process does not specify the road to success. Natural selection has the capacity of producing an incredible range of organisms, from the most asocial and competitive to the kindest and gentlest. The same process may not have specified our moral rules and values, but it has provided us with the psychological makeup, tendencies, and abilities to develop a compass for life's choices that takes the interests of the entire community into account, which is the essence of human morality.

Appendix A

Anthropomorphism and Anthropodenial

Often, when human visitors walk up to the chimpanzees at the Yerkes Field Station, an adult female named Georgia (figure 8) hurries to the spigot to collect a mouthful of water before they arrive. She then casually mingles with the rest of the colony behind the mesh fence of their outdoor compound, and not even the best observer will notice anything unusual about her. If necessary, Georgia will wait minutes with closed lips until the visitors come near. Then there will be shrieks, laughs, jumps, and sometimes falls, when she suddenly sprays them.

This is not a mere "anecdote," as Georgia does this sort of thing predictably, and I have known quite a few other apes good at surprising naive people—and not just naive people. Hediger (1955), the great Swiss zoo biologist, recounts how even when he was fully prepared to meet the challenge, paying attention to the ape's every move, he nevertheless got drenched by an old chimpanzee with a lifetime of experience with this game.

Once, finding myself in a similar situation with Georgia

Figure 8 Georgia, our naughtiest chimpanzee, fascinated by her own reflection in the camera lens. Photograph by the author.

(i.e., aware that she had gone to the spigot and was sneaking
up on me), I looked her straight in the eyes and pointed my
finger at her warning, in Dutch, "I have seen you!" She immediately stepped away and let part of the water drop, swal-
lowing the rest. I certainly do not wish to claim that she understands Dutch, but she must have sensed that I knew what
she was up to, and that I was not going to be an easy target.

The curious situation in which scientists who work with
these fascinating animals find themselves is that they cannot
help but interpret many of their actions in human terms,
which then automatically provokes the wrath of philosophers and other scientists, many of whom work with domes-
tic rats, or pigeons, or with no animals at all. Unable to speak
from firsthand experience, these critics must feel confident

indeed when they discard accounts by primatologists as anthropomorphie, and explain how anthropomorphism is to be avoided.

Although no reports of spontaneous ambush tactics in rats have come to my attention, these animals could conceivably be trained with patient reinforcement to retain water in their mouth and stand amongst other rats. And if rats can learn to do so, what is the big deal? The message of the critics of anthropomorphism is something along the lines of "Georgia has no plan; Georgia does not know that she is tricking people; Georgia just learns things faster than a rat." Thus, instead of seeking the origin of Georgia's actions within herself, and attributing intentions to her, they propose to seek the origin in the environment and the way it shapes behavior. Rather than being the designer of her own disagreeable greeting ceremony, this ape fell victim to the irresistible rewards of human surprise and annoyance. Georgia is innocent!

But why let her off the hook that easily? Why would any human being who acts this way be scolded, arrested, or held accountable, whereas any animal, even of a species that resembles us so closely, is considered a mere passive instrument of stimulus-response contingencies? Inasmuch as the absence of intentionality is as difficult to prove as its presence, and inasmuch as no one has ever proven that animals differ fundamentally from people in this regard, it is hard to see the scientific basis for such contrasting assumptions. Surely, the origin of this dualism is to be found partly outside of science.

The dilemma faced by behavioral science today can be summarized as a choice between cognitive and evolutionary parsimony (de Waal 1991, 1999). *Cognitive parsimony* is the traditional canon of American Behaviorism. It tells us not to invoke higher mental capacities if we can explain a phenomenon with ones lower on the scale. This favors a simple explanation, such as conditioned behavior, over a more complex one, such as intentional deception. This sounds fair enough (but see Sober 1990). *Evolutionary parsimony*, on the other hand, considers shared phylogeny. It posits that if closely related species act the same, the underlying mental processes are probably the same, too. The alternative would be to assume the evolution of divergent processes that produce similar behavior, which seems a wildly uneconomic assumption for organisms with only a few million years of separate evolution. If we normally do not propose different causes for the same behavior in, say, dogs and wolves, why should we do so for humans and chimpanzees?

In short, the cherished principle of parsimony has taken on two faces. At the same time that we are supposed to favor low-level over high-level cognitive explanations, we also should not create a double standard according to which shared human and ape behavior is explained differently. If accounts of human behavior commonly invoke complex cognitive abilities—and they most certainly do (Michel 1991)—we must carefully consider whether these abilities are perhaps also present in apes. We do not need to jump to conclusions, but the possibility should at least be allowed on the table.

Even if the need for this intellectual breathing room is most urgently felt in relation to our primate relatives, it is neither limited to this taxonomic group nor to instances of complex cognition. Students of animal behavior are faced with a choice between classifying animals as automatons or granting them volition and information-processing capacities. Whereas one school warns against assuming things we cannot prove, another school warns against leaving out what may be there: even insects and fish come across to the human observer as internally driven, seeking, wanting systems with awareness of their surroundings. Descriptions that place animals closer to us than to machines adopt a language that we customarily use for human action. Inevitably, these descriptions sound anthropomorphic.

Obviously, if anthropomorphism is defined as the misattribution of human qualities to animals, no one wishes to be associated with it. But much of the time, a broader definition is employed, namely the description of animal behavior in human, hence intentionalistic, terms. Even though no anthropomorphism proponent would propose to apply such language uncritically, even the staunchest opponents of anthropomorphism do not deny its value as a heuristic tool. It is this use of anthropomorphism as a means to get at the truth, rather than as an end in itself, that distinguishes its use in science from that by the layperson. The ultimate goal of the anthropomorphizing scientist is emphatically not the most satisfactory projection of human feelings onto the animal, but testable ideas and replicable observations.

This requires great familiarity with the natural history and special traits of the species under investigation, and an effort to suppress the questionable assumption that animals feel and think like us. Someone who cannot imagine that ants taste good cannot successfully anthropomorphize the anteater. So, in order to have any heuristic value at all, our language must respect the peculiarities of a species while framing them in a way that strikes a chord in the human experience. Again, this is easier to achieve with animals close to us than with animals, such as dolphins or bats, that move through a different medium or perceive the world through different sensory systems. Appreciation of the diversity of *Umwelten* (von Uexküll 1909) in the animal kingdom remains one of the major challenges of the student of animal behavior.

The debate about the use and abuse of anthropomorphism, which for years was confined to a small academic circle, has recently been thrust into the spotlight by two books: Kennedy's (1992) *The New Anthropomorphism*, and Marshall Thomas's (1993) *The Hidden Life of Dogs*. Kennedy reiterates the dangers and pitfalls of assuming higher cognitive capacities than can be proven, thus defending cognitive parsimony. Marshall Thomas, on the other hand, does not make any bones about the anthropomorphic bias of her informal study of canine behavior. In her best-seller, the anthropologist lets virgin bitches "save" their virginity for future "husbands" (i.e., ignore sexual attentions prior to meeting a favorite male, p. 56), watches wolves set out for the hunt without "pitying themselves" (p. 39), and looks into her dogs' eyes during a vicious gang attack seeing "no anger, no fear, no threat, no show of aggression, just clarity and overwhelming determination" (p. 68).

There is quite a difference between the use of anthropomorphism for communicatory purposes or in order to generate hypotheses, and the sort of anthropomorphism that does little else than project human emotions and intentions onto animals without justification, explication, or investigation (Mitchell et al. 1997). The uncritical anthropomorphism of Marshall Thomas is precisely what has given the practice a bad name, and has led critics to oppose it in all of its forms and disguises. Rather than let them throw out the baby with the bathwater, however, the only question that needs to be answered is whether a certain dose of anthropomorphism, used in a critical fashion, helps or hurts the study of animal behavior. Is it something that, as Hebb (1946) already noted, allows us to make sense of animal behavior, and, as Cheney and Seyfarth (1990: 303) declared, "works" in that it increases the predictability of behavior? Or is it something that, as Kennedy (1992) and others argue, needs to be brought under control, almost like a disease, because it makes animals into humans?

While it is true that animals are not humans, it is equally true that humans are animals. Resistance to this simple yet undeniable truth is what underlies the resistance to anthropomorphism. I have characterized this resistance as *anthropodenial*, the *a priori* rejection of shared characteristics between humans and animals. Anthropodenial denotes willful blindness to the human-like characteristics of animals, or the animal-like characteristics of ourselves (de Waal 1999). It reflects a pre-Darwinian antipathy to the profound similarities between human and animal behavior (e.g., maternal care, sexual behavior, power seeking) noticed by anyone with an open mind.

The idea that these similarities require unitary explanations is anything but new. One of the first to advocate cross-specific explanatory uniformity was David Hume (1985 [1739]: 226), who formulated the following touchstone in *A Treatise of Human Nature*:

Tis from the resemblance of the external actions of animals to those we ourselves perform, that we judge their internal likewise to resemble ours; and the same principle of reasoning, carry'd one step farther, will make us conclude that since our internal actions resemble each other, the causes, from which they are deriv'd, must also be resembling. When any hypothesis, therefore, is advanc'd to explain a mental operation, which is common to men and beasts, we must apply the same hypothesis to both.

It is important to add that, in contrast to American behaviorists, who two centuries after Hume would accommodate animals and humans within a single framework by seriously downgrading human mental complexity and relegating consciousness to the domain of superstition (e.g., Watson 1930), Hume (1985 [1739]: 226) held animals in high esteem, writing that "no truth appears to me more evident than that beasts are endow'd with thought and reason as well as men."

Strictly speaking, one cannot boast a unified theory of all behavior, human and animal, while at the same time decrying anthropomorphism. After all, anthropomorphism assumes similar experiences in humans and animals, which is exactly what one would expect in case of shared underlying processes. The behaviorists' opposition to anthropomorphism probably came about because no sane person would take seriously their claim that internal mental operations in *our* species are a figment of the imagination. The masses refused to accept that their own behavior could be explained without considering thoughts, feelings, and intentions. Don't we have mental lives, don't we look into the future, aren't we rational beings? Eventually, the behaviorists relented, exempting the bipedal ape from their theory of everything.

This is where the problem for other animals began. Once cognitive complexity was admitted in humans, the rest of the animal kingdom became the sole light-bearer of Behaviorism. Animals were expected to follow the law of effect to the absolute letter, and anyone who thought differently was just being anthropomorphic. Attribution of human-like experiences to animals was declared a cardinal sin. From a unified science, Behaviorism had deteriorated into a dichotomous one with two separate languages: one for human behavior, another one for animal behavior.

So, the answer to the question "Isn't anthropomorphism dangerous?" is that, yes, it is dangerous to those who wish to uphold a wall between humans and other animals. It places all animals, including humans, on the same explanatory plane. It is hardly dangerous, though, to those working from an evolutionarily perspective so long as they treat anthropomorphic explanations as hypotheses (Burghardt 1985). Anthropomorphism is a possibility among many, but one to be taken seriously given that it applies intuitions about ourselves to creatures very much like us. It is the application of human self-knowledge to animal behavior. What could be wrong with that? We apply human intuition in mathematics and chemistry, so why should we suppress it in the study of animal behavior? Stronger yet: does anyone truly believe that anthropomorphism is avoidable (Cenami Spada 1997)?

In the end we must ask: What kind of risk are we willing to take, the risk of underestimating animal mental life or the risk of overestimating it? There is a symmetry between anthropomorphism and anthropodenial, and since each has its strengths and weaknesses, there is no simple answer. But from an evolutionary perspective, Georgia's mischief is most parsimoniously explained in the same way we explain our own behavior—as the result of a complex, and familiar, inner life.

Appendix B

Do Apes Have a Theory of Mind?

I ntersubjectivity research on primates began with Menzel (1974), who released young chimpanzees into a large outdoor enclosure with only one of them aware of the location where food (or a toy snake) was hidden, while his or her companions possessed no such knowledge. The companions seemed perfectly capable, however, of "guessing" what to expect based on the behavior of the "knower." Menzel's classic experiment, combined with Humphrey's (1978) notion of animals as "natural psychologists" and Premack and Woodruff's (1978) "theory of mind" (ToM), inspired the guesser-versus-knower paradigm still popular today in intersubjectivity research on both apes and children.

ToM refers to the ability to recognize the mental states of others. If you and I meet at a party, and I believe that you believe that we have never met before (even though we did meet), I have a theory about what is going on in your head. Given that some scientists claim this ability to be uniquely human, it is ironic that the whole ToM concept originated with primate research. It has had serious ups and downs since. Some concluded from failed demonstrations that apes must lack ToM (e.g., Tomasello 1999; Povinelli 2000). Negative results are impossible to interpret, though. As the saying goes: Absence of evidence is no evidence of absence. An experiment may not work for reasons that have nothing to do with the capacity in question. When comparing apes and children, for example, one problem is that the experimenter is invariably human, so that only the apes face a species barrier (de Waal 1996).

To captive apes, we must come across as all-mighty and all-knowing. We approach chimpanzees in our care while having heard the latest news about them from others (e.g., we have been called on the telephone about a new injury or birth). The chimpanzees must notice that we often know things before we have seen them. This makes humans inherently unsuitable as participants in experiments on the role of seeing in knowing, a centerpiece of ToM research.

All that most experiments have done thus far is test the ape's theory of the human mind. We would do better to focus on the ape's theory of the ape mind. When the human experimenter is cut out of the picture, chimpanzees seem to realize that if another has seen hidden food, this individual knows (Hare et al. 2001). This finding, along with growing evidence for visual perspective-taking in apes (Shillito et al. 2005; Bräuer et al.; 2005; Hare et al., in press; Hirata 2006), has thrown the question of animal ToM wide open again. In an unexpected twist (because the debate revolves around humans and apes), a capuchin monkey at the University of Kyoto recently passed a number of seeing-knowing tests with flying colors (Kuroshima et al. 2003). A few such positive outcomes suffice to question all previous negative ones.

The only way to get to the bottom of ape intelligence is to

design experiments that engage them intellectually and emotionally. Rescuing an infant from an attack, outmaneuvering a rival, avoiding conflict with the dominant male, sneaking away with a mate, are the kinds of problems apes are good at solving. There are many reports suggesting ToM in ape social life, and even though these are usually single events—sometimes disparagingly labeled "anecdotes"—I consider them extremely meaningful. After all, one step by one man on the moon has been sufficient for us to claim that going there is within our ability. If an experienced, reliable observer reports a remarkable incident, science had better pay close attention. At the very least, such reports can be of heuristic value (de Waal 1991). With regard to apes taking another's perspective, we do not just have a few stories, but a great many of them. In the lead text I have related the stories of Kuni and the bird and Jakie and his aunt. Let me offer two more examples from de Waal (1989a).

> The two-meter-deep moat in front of the old bonobo enclosure at the San Diego Zoo had been drained for cleaning. After having scrubbed the moat and released the apes, the keepers went to turn on the valve to refill it with water when all of a sudden the old male, Kakowet, came to their window, screaming and frantically waving his arms so as to catch their attention. After so many years, he was familiar with the cleaning routine. As it turned out, several young bonobos had entered the dry moat but were unable to get out. The keepers provided a ladder. All bonobos got out except for the smallest one, who was pulled up by Kakowet himself.
>
> This story matches another moat observation made at the same enclosure a decade later. By this time, the zoo had wisely decided against water in the moat as apes cannot

swim. A chain hung permanently down into it, and the bonobos visited the moat whenever they wanted. If the alpha male, Vernon, disappeared into the moat, however, a younger male, Kalind, sometimes quickly pulled up the chain. He would then look down at Vernon with an open-mouthed play face while slapping the side of the moat. This expression is the equivalent of human laughter: Kalind was making fun of the boss. On several occasions, the only other adult, Loretta, rushed to the scene to rescue her mate by dropping the chain back down, and standing guard until he had gotten out.

Both observations tell us something about perspective-taking. Kakowet seemed to realize that filling the moat while the juveniles were still in it wouldn't be a good idea even though this would obviously not have affected him. Both Kalind and Loretta seemed to know what purpose the chain served for someone at the bottom of the moat and to act accordingly, the one by teasing, the other by assisting the dependent party.

I am personally convinced that apes take one another's perspective, and that the evolutionary origin of this ability is not to be sought in social competition, even if it is readily applied in this domain (Hare and Tomasello 2004), but in the need for cooperation. At the core of perspective-taking is emotional linkage between individuals—widespread in social mammals—upon which evolution (or development) builds ever more complex manifestations, including appraisal of another's knowledge and intentions (de Waal 2003).

Because of this possible connection between empathy and ToM, the bonobo is a crucial species for further research, as it may be the most empathic ape (de Waal 1997a). Recent DNA comparisons show that humans and bonobos share a microsatellite related to sociality that is absent in the chimpanzee (Hammock and Young 2005). This may be insufficient to decide which of our two closest relatives, the bonobo or the chimpanzee, most resembles the last common ancestor, but it definitely calls for close attention to the bonobo as model of human social behavior.

Appendix C

Animal Rights

𝒟

After a narrow escape, the Thompson gazelle calls her lawyer, complaining that her freedom to graze wherever she wants has once again been violated. Should she sue the cheetah, or does the lawyer perhaps feel that predators have rights, too?

Absurd, of course, and I certainly applaud efforts to prevent animal abuse, but I do have serious questions about the approach that now has led American law schools to start offering courses in "animal law." What they mean is not the law of the jungle, but the extension of principles of justice to animals. Animals are not mere property, according to some, like Steven M. Wise, the lawyer teaching the course at Harvard. They deserve rights as solid and uncontestable as the constitutional rights of people. Some animal rights lawyers have even argued that chimpanzees deserve the right to bodily integrity and liberty.

This view has gained some currency. For instance, the Court of Appeals for the District of Columbia gave a human zoo visitor the right to sue to get companionship for chimpanzees. And over the last decade, state legislatures have upgraded animal cruelty crimes to felonies from misdemeanors.

The debate over animal rights is not new. I still remember some surrealistic debates among scientists in the 1970s that dismissed animal suffering as a bleeding-heart issue. Amid stern warnings against anthropomorphism, the then prevailing view was that animals were mere robots, devoid of feelings, thoughts, or emotions. With straight faces, scientists would argue that animals cannot suffer, at least not the way we do. A fish is pulled out of the water with a big hook in its mouth, it thrashes around on dry land, but how could we possibly know what it feels? Isn't all of this pure projection?

This thinking changed in the 1980s with the advent of cognitive approaches to animal behavior. We now use terms like "planning" and "awareness" in relation to animals. They are believed to understand the effects of their own actions, to communicate emotions and make decisions. Some animals, like chimpanzees, are even considered to have rudimentary politics and culture.

In my own experience, chimpanzees pursue power as relentlessly as some people in Washington and keep track of given and received services in a marketplace of exchange. Their feelings may range from gratitude for political support to outrage if one of them violates a social rule. All of this goes far beyond mere fear, pain, and anger: the emotional life of these animals is much closer to ours than once held possible.

This new understanding may change our attitude toward chimpanzees and, by extension, other animals, but it remains a big leap to say that the only way to ensure their decent treatment is to give them rights and lawyers. Doing so is the American way, I guess, but rights are part of a social contract that makes no sense without responsibilities. This is the reason that the animal rights movement's outrageous parallel with the abolition of slavery—apart from being insulting—is morally flawed: slaves can and should become full members of society; animals cannot and will not.

Indeed, giving animals rights relies entirely upon our good will. Consequently, animals will have only those rights that we can handle. One won't hear much about the rights of rodents to take over our homes, of starlings to raid cherry trees, or of dogs to decide their owner's walking route. Rights selectively granted are, in my book, no rights at all.

What if we drop all this talk of rights and instead advocate a sense of obligation? In the same way that we teach children to respect a tree by mentioning its age, we should use the new insights into animals' mental life to foster in humans an ethic of caring in which our interests are not the only ones in the balance.

Even though many social animals have evolved affectionate and altruistic tendencies, they rarely if ever direct these to other species. The way the cheetah treats the gazelle is typical. We are the first to apply tendencies that evolved within the group to a wider circle of humanity, and could do the same to other animals, making care, not rights, the centerpiece of our attitude.

APE RETIREMENT

The discussion above (modified from an Op-Ed piece that appeared in the *New York Times* of August 20, 1999, under the title "We the People [and Other Animals] . . .") questions the "rights" approach, but fails to indicate how I feel about invasive medical research.

The issue is complex, because I believe that our first moral obligation is to members of our own species. I know of no animal rights advocate in need of urgent medical attention who has refused such attention. This is so even though all modern medical treatments derive from animal research: anyone who walks into a hospital makes use of animal research then and there. There seems a consensus, therefore, even among those who protest animal testing, that human health and well-being take priority over almost anything else. The question then becomes: What are we willing to sacrifice for it? What kind of animals are we willing to subject to invasive medical studies, and what are the limits on the procedures? For most people, this is a matter of degree, not of absolutes. Using mice to develop new cancer drugs is not put at the same level as shooting pigs to test the impact of bullets, and shooting pigs is not put at the same level as giving a lethal disease to a chimpanzee. In a complex gain-versus-pain calculation, we decide on the ethics of animal research based on how we feel about procedures, animal species, and human benefits.

Without going into the reasons and incongruities of why we favor some animals over others and some procedures over others, I do personally believe that apes deserve special status. They are our closest relatives with very similar social and emotional lives and similar intelligence. This is, of course, an anthropocentric argument if there ever was one, but one shared by many people familiar with apes. Their closeness to us makes them both ideal medical models and ethically problematic ones.

Although many people favor a logical moral stance, based on straightforward empirical facts (such as the oft-mentioned ability of apes to recognize themselves in a mirror), no reasoned moral position seems airtight. I believe in the emotional basis of moral decisions, and since empathy with creatures that bodily and psychologically resemble us comes easily to us, apes mobilize in us more guilty feelings about hurting them than do other animals. These feelings play a role when we decide on the ethics of animal testing.

Over the years, I have seen the prevailing attitude shift from emphasis on the medical usefulness of apes to emphasis on their ethical status. We now have reached the point that they are medical models of last resort. Any medical study that can be done on monkeys, such as baboons or macaques, will not be permitted on chimpanzees. Since the number of ape-specific research questions is dwindling, we are facing a "surplus" of chimpanzees. This is the medical community's way of saying that we now have more chimpanzees than needed for medical research.

I consider this a positive development, and am all for it progressing further until chimpanzees can be phased out completely. We have not reached this point yet, but increasing reluctance to use chimpanzees has led the National Institutes of Health to take the historic step of sponsoring retirement for these animals. The most important facility is Chimp Haven (www.chimphaven.org), which in 2005 opened a large outdoor facility to retire chimpanzees taken off medical protocols.

In the meantime, apes will remain available for noninvasive studies, such as those on aging, genetics, brain imaging, social behavior, and intelligence. These studies do not require harming the animals. The shorthand definition that

I use for noninvasive research is "the sort of research we wouldn't mind doing on human volunteers." This would mean no testing of compounds on them, nor giving them any disease they don't already have, no disabling surgeries, and so on.

Such research will help us continue to learn about our closest relatives in nonstressful, even pleasant, ways. I add the latter, because the chimpanzees I work with are keen on computerized testing: the easiest way to get them to enter our testing facility is to show them the cart with the computer on top. They rush into the doors for an hour of what they see as games and what we see as cognitive testing.

Ideally, all research on apes should be mutually beneficial and enjoyable.

PART II

COMMENTS

The Uses of Anthropomorphism

ROBERT WRIGHT

rans de Waal's carefully documented and richly descriptive accounts of nonhuman primate social behavior have contributed vastly to our understanding of
both nonhuman primates and human ones. One thing that
has made his accounts so intellectually stimulating is his
willingness to use provocatively anthropomorphic language
in analyzing the behavior and mentality of chimpanzees and
other nonhuman primates. Not surprisingly, he has drawn
some criticism for this anthropomorphism. Almost invariably, I think, the criticism is misguided. However, while convinced of the value of his use of anthropomorphic language,
I do believe that de Waal is occasionally uncritical in the *kind*
of anthropomorphic language he uses.

I'd like to first flesh this point out and then argue that
one benefit of fleshing it out is to expand our perspective
on human morality. In particular: Clarifying the question
of what kind of anthropomorphic language is appropriate
for chimpanzees, our nearest relatives, sheds light on de
Waal's distinction between a "naturalistic" theory of human

morality and a "veneer" theory of human morality—between the idea that morality has a firm foundation in the genes, and the idea that what we call "morality" is a mere "cultural overlay," and often takes the form of a kind of moral posturing that masks an amoral if not immoral human nature. I think de Waal misunderstands the perspective of some people he labels "veneer theorists" (me, for example) and as a result misses something important and edifying that evolutionary psychology can bring to discussions about human morality, namely: evolutionary psychology suggests the value of a third kind of theory about human morality that—to adapt de Waal's terminology—we might call "naturalistic veneer theory." This third alternative will be easier to understand once we've pondered the question of what kind of anthropomorphic language is appropriate for chimpanzees, the question to which I'll now turn.

TWO KINDS OF ANTHROPOMORPHIC LANGUAGE

It is almost impossible to read de Waal's great book *Chimpanzee Politics* without being struck by the behavioral parallels between chimpanzees and humans. For example: In both species social status brings tangible rewards, in both species individuals seek it, and in both species individuals form social alliances that help them seek it. Given the close evolutionary relationship between human beings and chimpanzees, it is certainly plausible that these external behavioral parallels are matched by internal parallels—that is, that there is some inter-species commonality in the biochemical mechanisms governing the behavior and in the corresponding subjective experience. Facial expressions, gestures, and postures that accompany certain chimpanzee behaviors certainly reinforce this conjecture.

But what is the exact nature of these commonalities? What particular subjective experiences, for example, might we share with chimpanzees? Here is where I take issue with an interpretive tendency of de Waal's.

There are two broad categories of anthropomorphic language. First, there is *emotional* language: We can say that chimpanzees feel compassionate, outraged, aggrieved, insecure, et cetera. Second, there is *cognitive* language, language that attributes conscious knowledge and/or reasoning to animals: We can say that chimpanzees remember, anticipate, plan, strategize, et cetera.

It isn't always clear from the behavioral evidence alone which kind of anthropomorphic language is in order. Fairly often, in both humans and nonhuman primates, a behavior could in principle be explained either as a product of conscious reflection and strategizing or as a product of essentially emotional reaction.

Consider "reciprocal altruism." In the case of both humans and chimpanzees we see what looks at the behavioral level like reciprocal altruism. That is, individuals strike up regular relationships with other individuals that feature the giving of goods such as food or the giving of services such as social support; and the giving is somewhat symmetrical over time: I scratch your back, you scratch mine.

In the case of humans, we know through introspection that these relationships of mutual support can be governed at either of two levels—at the cognitive level or at the emotional level. (In real life there is typically a mixture of cognitive and emotional factors, but usually one predominates, and in any event I'll consider "pure" examples of each for purposes of clarity in the thought experiment that follows.)

Consider two scholars who work in the same field but have never met. Suppose you are one of the scholars. You are writing a paper that offers you an opportunity to cite the other scholar. The citation isn't essential; the paper would be fine without it. But you think to yourself, "Well, maybe if I cite this person, this person will cite me down the road, and this might lead to a pattern of mutual citation that would be good for both of us." So you cite this scholar, and the stable relationship of mutual citation that you anticipated—a kind of "reciprocal altruism"—indeed ensues.

Now imagine an alternative path to the same outcome. While working on your paper, you meet this scholar at a conference. You immediately hit it off, warming to each other as you discuss your common intellectual interests and opinions. Later, while finishing the paper, you cite this scholar out of sheer friendship; you don't so much decide to cite him/her as *feel* like citing him/her. He/she later cites you, and a pattern of mutual citation, of "reciprocal altruism," ensues.

In the first case, the relationship of mutual citation feels like a result of strategic calculation. In the second case, it feels more like a case of simple friendship. But to the outside observer—someone who is just observing the tendency of these two scholars to cite each other—it is hard to distinguish between the two kinds of motivation. It is hard to say whether the pattern of mutual citation is driven more by strategic calculation or by friendship, because either of those dynamics can in principle lead to the observed outcome: a stable relationship of mutual citation.

Suppose the outside observer is now given an additional

piece of information: these two scholars not only tend to cite each other; they tend to be on the same side of the great, divisive issues in their field. Alas, this doesn't help much either, because both of the dynamics in question—strategic calculation and friendly feeling—are known to lead to this specific outcome: not just mutual citation, but mutual citation between intellectual allies. After all, (a) if you're consciously choosing a partner in reciprocal citation, you'll be inclined to choose someone who shares your strategic interests, namely the advancement of your position on major intellectual issues; (b) if you're operating instead on the basis of friendly feelings, you're still likely to wind up paired with an intellectual ally, since one of the primary contributors to friendly feelings is agreement on contentious issues.

That the guidance of emotions—of "friendly feelings"—can lead to the same outcome as the guidance of strategic calculation is no coincidence. According to evolutionary psychology, human emotions were "designed" by natural selection to serve the strategic interests of individual human beings (or, more precisely, to further the proliferation of the individual's genes in the environment of our evolution—but for purposes of this discussion we can assume the interests of the individual and of the individual's genes align, as they often do). In the case of friendly feelings, we are "designed" to warm up to people who share our opinions on contentious issues because, during evolution, these are people it would have been advantageous to form alliances with.

This is the generic reason that it is often hard for an outside observer to say whether a given human behavior was driven more by strategic calculation or by emotions: *because many emotions are proxies for strategic calculation.* (As for why natural selection created these proxies for strategic calculation:

these emotions evolved, presumably, either before our ancestors were very good at conscious strategic calculation or in cases where conscious awareness of the strategy being pursued was disadvantageous.)

WHAT IS IT LIKE TO BE A CHIMP?

Now, with this thought experiment in hand, we can return to the question of anthropomorphic language—in particular the question of when "emotional" anthropomorphism is in order and when "cognitive" anthropomorphism is in order. In analyzing chimpanzee dynamics, and trying to decide whether chimps are engaging in conscious calculation or simply being guided by emotions, we face the same difficulty we faced with the two scholars: because the emotions in question were "designed" by natural selection to yield strategically effective behavior, emotionally driven behaviors and consciously calculated behaviors may look the same to the outside observer.

For example, if two chimpanzees are both frozen out of the power structure—that is, they are not part of the coalition that keeps the alpha male in power, and so don't partake of the resources that the alpha shares with coalition partners—then they may form an alliance that challenges the alpha. But it's hard to say whether the initial formation of the alliance is a product of conscious strategic calculation or merely of "friendly feelings" that were "designed" by natural selection as proxies for conscious strategic calculation. Therefore it's hard to choose between "cognitive" anthropomorphic language ("the chimps saw that they shared a strategic interest and decided to form an alliance") and "emotional" anthropomorphic language ("the chimps, upon dimly sensing their shared plight, developed friendly feelings, and attendant feelings of mutual obligation, that drew them into alliance.")

In such ambiguous cases, de Waal seems to have a tendency to favor cognitive over emotional anthropomorphic language. One example from *Chimpanzee Politics* involves the alpha male Yeroen and a lower status chimp, Luit, who has in the past accepted his subordinate status but will soon mount a challenge to Yeroen's alpha position by initiating a fight. De Waal observes that, during the period leading up to the challenge, Yeroen began consolidating social bonds, notably by increasing the time he spent grooming females and otherwise interacting with them. From this de Waal infers that Yeroen "already sensed that Luit's attitude was changing and he knew that his position was threatened."[1]

Yeroen presumably did in some sense or another "sense" a change of attitude, and this may well account for his sudden interest in the politically pivotal females. But must we assume, with de Waal, that Yeroen "knew" about—consciously anticipated—the coming challenge and decided on measures to head it off? Or mightn't Luit's growing assertiveness simply have inspired pangs of insecurity that pulled Yeroen into closer touch with his friends?

Certainly genes encouraging an unconsciously rational response to threats could in theory thrive via natural selection. When a baby chimp or a baby human, sighting a scary animal, retreats to its mother, the response is logical, but the youngster presumably isn't conscious of the logic. Or, to take an example more analogous to the Yeroen-Luit example: if a

---

[1] De Waal (1982), *Chimpanzee Politics*, Baltimore, MD: Johns Hopkins University Press, p. 98.

human being is treated with unexpected disrespect by some acquaintances, he or she may be filled with a feeling of insecurity and therefore, upon subsequently encountering a friend or relative, reach out to that friend or relative more than usual, and, upon getting positive feedback, feel more warmly toward that friend or relative than usual. Here "insecurity" is a proxy emotion for strategic calculation; it encourages us to strengthen our bonds with allies after encountering social antagonism.

A more sweeping example of de Waal's seeming preference for cognitive over emotional anthropomorphism comes when he refers to Luit's "policy reversals, rational decisions and opportunism" and then asserts, "there is no room in this policy for sympathy and antipathy."[2] Actually, many of Luit's policy reversals, and much of his opportunism, can in principle be explained *in terms of* sympathy and antipathy: he feels sympathy toward chimps when strategic interests dictate alliance with them and he feels antipathy toward chimps when strategic interests dictate conflict with them or indifference toward them. Any human being is familiar with how rapidly feelings toward another human being can swing between sympathy and antipathy; and any deeply introspective human being would have to admit that sometimes these swings have a certain strategic convenience about them.

Of course, subjective experience being intrinsically private, it is hard to say for sure that de Waal is wrong—that the strategic behaviors in question are more under emotional than under cognitive guidance. But here are some interrelated considerations suggesting that this is the case:

[2] De Waal (1982), p. 196.

1) It is a cogent surmise, for various reasons, that in the primate lineage the emotional governance of behavior has preceded, in evolutionary time, the consciously strategic guidance of behavior. (One reason for this surmise is the relative evolutionary age of parts of the human brain associated with emotions, on the one hand, and with planning and reasoning, on the other. Also notable is the prominence of these respective parts of the brain compared with their prominence in nonhuman primates—e.g., the prominence of the human frontal lobes, associated with planning and reasoning.)

2) Given that human beings, though manifestly capable of conscious strategizing, nonetheless have emotions that encourage strategically sound behaviors, it seems likely that our near relatives the chimpanzees, who exhibit analogous strategically sound behaviors, also have such emotions.

3) If indeed chimpanzees have emotions that could generate strategically sound behaviors, one has to ask why natural selection would add a second, functionally redundant layer of guidance (conscious strategy). Of course, in the case of human beings evolution *did* ultimately supplement emotional guidance with cognitive guidance. But when we speculate as to why this is the case, we tend to cite factors that don't seem to apply to chimpanzees (e.g., humans have complex language and use it to discuss strategic plans with allies, or to explain why they did things, etc.).

For these reasons, when dealing with nonhuman primates, I would propose a bias that is the opposite of the bias de Waal seems to employ. In cases where either emotional guidance or consciously strategic guidance could in principle explain the behavior, I would favor emotional guidance as the tentatively preferred explanation, pending further data. That is: All other things being equal, I would favor emotional anthropomorphic language over cognitive anthropomorphic language when dealing with nonhuman primates.

You might call this the *principle of anthropomorphic parsimony*. One reason I consider it parsimonious is that it involves the use of only one type of anthropomorphic language (emotional) whereas the alternative favored by de Waal, though ostensibly involving only one type of anthropomorphic language (cognitive), implicitly involves both kinds. After all, it seems very likely that, if chimpanzees indeed have the capacity for extensive conscious strategizing, as de Waal believes, they also have a parallel and intertwined system of emotional proxies for strategic calculation—since, after all, that is the case with the one primate species *known* to have the capacity for extensive conscious strategizing (us), a species, moreover, that is closely related to chimpanzees. Assuming this is the case—that a close relative of humans that had extensive conscious strategizing abilities would also have intertwined emotional proxies for strategizing—then to attribute conscious strategizing to chimpanzees is to implicitly attribute both conscious strategizing and some degree of emotional guidance to them. And in cases where emotional guidance alone would in theory suffice for explanatory purposes, this implied attribution of both cognitive and emotional guidance is the less parsimonious alternative.

AN EXTRA-SCIENTIFIC CONSIDERATION

Though I consider this proposed rule for the use of anthropomorphic language desirable on scientific grounds—on grounds of parsimony—I should acknowledge that there is a second reason for my attraction to it: it encourages a perspective on human behavior that can be morally enriching. Appreciating how emotions can lead to strategically sophisticated behavior in chimpanzees helps us appreciate that we human beings may be more in the thrall of emotional guidance than we realize. In particular: our moral judgments are subtly and pervasively colored by emotionally mediated self-interest.

To clarify this point, let me back up and approach the subject of human morality from another angle: in terms of the distinction de Waal made in his first Tanner lecture between a "veneer" theory of morality and a "naturalistic" theory. Veneer Theory holds that human morality is a thin "cultural overlay" that hides an amoral if not immoral human nature. The alternative, "naturalistic" theory, as I understand it, holds that our moral impulses are rooted in our genes— and that we are therefore to a considerable extent, as the title of one of de Waal's books has it, "good natured."

De Waal classifies me as a "veneer theorist" on the basis of my book *The Moral Animal*. Let me briefly argue that I don't belong in this category, and in the process argue that his dichotomy between "veneer" theory and a "naturalistic" theory is perhaps too simple, omitting a third theoretical category in which I do belong. Then I can argue that using emotionally anthropomorphic language to describe chimpanzee behavior helps illuminate this third theoretical perspective, and that viewing human behavior from this third vantage point has edifying effects.

In *The Moral Animal*, far from describing morality as a "cultural overlay," I in fact argue that various impulses and behaviors commonly described as moral are grounded in our genes. Kin-selected altruism is one example. Another example is the sense of justice—the intuition that good deeds should be rewarded and bad deeds should be punished; de Waal's work, in fact, helped convince me that a rudimentary (and, I would say, heavily emotional) version of this intuition is probably present in chimpanzees, and that in both chimps and humans the intuition is a product of the evolutionary dynamic of reciprocal altruism.

These features of human nature, grounded in the genes, are often exercised in what I would call truly moral fashion. (That is—to adopt a crude and somewhat utilitarian version of the Kantian litmus test that Christine Korsgaard explicates in her paper—the world is a better place to the extent that the behaviors generated by these features are generated under comparable circumstances by human beings in general.) So I don't think I deserve, in a general way, the label de Waal gives me—a "veneer theorist" who considers morality a "cultural overlay."

To be sure, I do believe that some of our genetically based moral intuitions are (*sometimes*) subject to subtle biases that steer them away from the truly moral. But even here I'm not conforming to the archetype of the "veneer theorist," because I believe these biases to be themselves grounded in the genes, not mere "cultural overlay." For example, in deciding how to exercise the sense of justice—in deciding who has done good and who has done bad, whose grievances are valid and whose aren't—humans seem naturally to pass judgments that work in favor of family and friends and against enemies and rivals. This is one reason I don't agree with de Waal's apparent position that we are in some fairly general sense "good natured"—a view he seems to associate with "naturalistic theory."

Rather, I belong to a third category. I believe (a) that the human moral "infrastructure"—the part of human nature that we draw on for moral guidance, and that includes some specific moral intuitions—is genetically rooted, not a "cultural overlay"; but (b) this infrastructure is not infrequently subject to systematic "corruption" (i.e., departure from what I would call true morality) that is itself rooted in the genes (and is so rooted because it served the Darwinian interests of our ancestors during evolution).

In this view, our moral judgments, though reached through a seemingly conscious and rational process of deliberation—a cognitive process—can be biased subtly by emotional factors. For example, an only semiconscious undercurrent of hostility toward a rival may bias our judgment as to whether he is guilty of some crime, even though we convince ourselves that we have weighed the evidence objectively. We may honestly believe that our opinion that he deserves, say, the death penalty is a product of pure cognition, with no emotional influence; but the emotional influence can in fact be decisive, and was "designed" by natural selection to be that way.

My own view is that if everyone were more aware of the ways emotion subtly biases their moral judgments, the world would be a better place, because we would be less likely to comply with these morally corrupting biases. I thus see some virtue in anything that makes people more self-aware in this regard. And I think using emotionally anthropomorphic language to describe certain aspects of chimpanzee social life—in addition to being defensible on sheerly scientific grounds—can have this effect. For seeing how subtly but powerfully emotions can guide the behavior of chimpanzees helps us see how subtly but powerfully emotions can influence our own behavior, including behaviors that we like to think of as products of pure reason.

To put the matter another way: When we see chimpanzees behaving in a strikingly human manner, we can describe the parallel in at least two ways. We can say, "Gosh, chimpanzees are more impressive than I'd thought"—a conclusion we're especially likely to reach if we see their behavior as cognitively driven. Or, alternatively, we can say, "Gosh, humans are less exalted than I'd thought"—a conclusion we're especially likely to reach if we see that relatively simple and ancient emotions can yield seemingly sophisticated behaviors in chimpanzees and hence, presumably, in humans. The latter conclusion is, in addition to being valid, edifying.

I want to stress, in closing, that most of the anthropomorphic language de Waal uses in *Chimpanzee Politics* and elsewhere I have no quarrel with (such as his speculatively attributing a sense of "honor"—i.e., something like pride—to chimpanzees). Still, I do think the two examples I've cited are telling, and that they are not wholly unrelated to his (too simple, in my view) dichotomy between a "veneer" theory of morality and a "naturalistic" theory of morality. Appreciating how subtly and powerfully emotions can influence behavior is the first step, I think, toward appreciating the existence of, and the importance of, the third theoretical category I've outlined.

I'm tempted to call this third theoretical orientation "naturalistic veneer theory," since it does see humans as often covering self-serving motives with a moralistic veneer, but sees the veneer-building process itself as genetically, not just culturally, grounded. That label has the shortcoming of not conveying that a good many of our natural moral impulses do have truly moral consequences (by my lights, at least). Still, combining "naturalistic" and "veneer" gets us closer to the truth, in this context, than leaving these words by themselves.

Morality and the Distinctiveness of Human Action

CHRISTINE M. KORSGAARD

What is different about the way we *act* that makes us, and not any other species, moral beings?
    —Frans de Waal[1]

A moral being is one who is capable of comparing his past and future actions or motives, and of approving or disapproving of them. We have no reason to suppose that any of the lower animals have this capacity.
    —Charles Darwin[2]

T wo issues confront us. One concerns the truth or falsehood of what Frans de Waal calls "Veneer Theory." This is the theory that morality is a thin veneer

[1] In *Good Natured: The Origins of Right and Wrong in Humans and Other Animals* Cambridge, MA: Harvard University Press, 1996, p. 111.
[2] In *The Descent of Man, and Selection in Relation to Sex* (1871), Princeton: Princeton University Press, 1981, pp. 88–89.

on an essentially amoral human nature. According to Veneer Theory, we are ruthlessly self-interested creatures, who conform to moral norms only to avoid punishment or disapproval, only when others are watching us, or only when our commitment to these norms is not tested by strong temptation. The second concerns the question whether morality has its roots in our evolutionary past, or represents some sort of radical break with that past. De Waal proposes to address these two questions together, by adducing evidence that our closest relatives in the natural world exhibit tendencies that seem intimately related to morality—sympathy, empathy, sharing, conflict resolution, and so on. He concludes that the roots of morality can be found in the essentially social nature we share with the other intelligent primates, and that therefore morality itself is deeply rooted in our nature.

I begin with the first issue. Veneer Theory is, in my view, not very tempting. In philosophy, it is most naturally associated with a certain view of practical rationality and of how practical rationality is related to morality. According to this view, what it is rational to do, as well as what we naturally do, is to maximize the satisfaction of our own personal interests. Morality then enters the scene as a set of rules that constrain this maximizing activity. These rules may be based on what promotes the common good, rather than the individual's good. Or they may, as in deontological theories, be based on other considerations—justice, fairness, rights, or what have you. In either case, Veneer Theory holds that these constraints, which oppose our natural and rational tendency to pursue what is best for ourselves, and which are therefore unnatural, are all too easily broken through. De Waal seems to accept the idea that it is rational to pursue your own best interests, but wants to reject the associated view that morality is unnatural, and therefore he tends to favor an emotion-based or sentimentalist theory of morality.

There are a number of problems with Veneer Theory. In the first place, despite its popularity in the social sciences, the credentials of the principle of pursuing your own best interests as a principle of practical reason have never been established. To show that this is a principle of practical reason one would have to demonstrate its normative foundation. I can think of only a few philosophers—Joseph Butler, Henry Sidgwick, Thomas Nagel, and Derek Parfit among them—who have even attempted anything along these lines.[3] And the idea that what people actually do is pursue their own best interests is, as Butler pointed out long ago, rather laughable.[4]

In the second place, it is not even clear that the idea of self-interest is a well-formed concept when applied to an animal as richly social as a human being. Unquestionably, we have some irreducibly private interests—in the satisfaction of our appetites, in food and a certain kind of sex, say. But our personal interests are not limited to *having* things. We

---

[3] Butler, in *Fifteen Sermons Preached at the Rolls Chapel* (1726), partly reprinted in *Five Sermons Preached at the Rolls Chapel and A Dissertation Upon the Nature of Virtue*, edited by Stephen Darwall, Indianapolis: Hackett Publishing Company, 1983; Sidgwick, in *The Methods of Ethics (1st ed., 1874, 7th ed., 1907)*. Indianapolis: Hackett Publishing Company, 1981); Nagel, in *The Possibility of Altruism* (Princeton: Princeton University Press, 1970); and Parfit, in *Reasons and Persons* (Oxford: Clarendon Press, 1984). For a discussion of the problems with providing a normative foundation for this supposed rational principle, see my "The Myth of Egoism" published by the University of Kansas as the Lindley Lecture for 1999.

[4] "Men daily, hourly sacrifice the greatest known interest to fancy, inquisitiveness, love, or hatred, any vagrant inclination. The thing to be lamented is not that men have so great a regard to their own good or interest in the present world, for they have not enough, but that they have so little to the good of others." Butler, *Five Sermons Preached at the Rolls Chapel and A Dissertation Upon the Nature of Virtue*, p. 21.

also have interests in *doing* things and *being* things. Many of these interests cannot set us wholly against the interests of society, simply because they are unintelligible outside of society and the cultural traditions that society supports. You could intelligibly want to be the world's greatest ballerina, but you could not intelligibly want to be the world's only ballerina, since, at least arguably, if there were only one, there wouldn't be any. Even for having things there is a limit to the coherent pursuit of self-interest. If you had all the money in the world, you would not be rich. And of course we also have genuine interests in certain other people, from whom our own interests cannot be separated. So the idea that we can clearly identify our own interests as something set apart from or over against the interests of others is strained to say the least.

And yet even this is not the deepest thing wrong with Veneer Theory. Morality is not just a set of obstructions to the pursuit of our interests. Moral standards define ways of relating to people that most of us, most of the time, find natural and welcome. According to Kant, morality demands that we treat other people as ends in themselves, never merely as means to our own ends. Certainly we do not manage to treat all other people at all times in accordance with this standard. But the image of someone who never treated *anyone* else as an end in himself and never expected to be treated that way in return is even more unrecognizable than that of someone who always does so. For what we are then imagining is someone who always treats *everyone else* as a tool or an obstacle and always expects to be treated that way in return. What we are imagining is someone who never spontaneously and unthinkingly tells the truth in ordinary conversation, but constantly calculates the effects of what he says to others on the promotion of his projects. What we are imagining is someone who doesn't resent it (though he dislikes it) when he himself is lied to, trampled on, and disregarded, because deep down he thinks that is all that one human being really has any reason to expect from any other. What we are imagining, then, is a creature who lives in a state of deep internal solitude, essentially regarding himself as the only person in a world of potentially useful things—although some of those things have mental and emotional lives and can talk or fight back.[5] It is absurd to suggest that this is what most human beings are like, or long to be like, beneath a thin veneer of restraint.

But it is also absurd to think that nonhuman animals are motivated by self-interest. The concept of what is in your own best interests, if it makes any sense at all, requires a kind of grip on the future and an ability to calculate that do not seem available to a nonhuman animal. Just as importantly, *acting* for the sake of your best interests requires the capacity to be *motivated* by the abstract conception of your overall or long-term good. The idea of self-interest seems simply out of place when thinking about nonhuman action. I am not at all inclined to deny that the other intelligent animals do things on purpose, but I would expect these purposes to be local and concrete—to eat something, mate with someone, avoid punishment, have some fun, stop the fight—but not to do what is best for themselves on the whole. Nonhuman animals are not self-interested. It seems more likely that they are, in Harry Frankfurt's phrase, wanton: they act on the instinct or desire or emotion that comes uppermost. Learning

---

[5] I owe some of these points to Thomas Nagel, *The Possibility of Altruism*, pp. 82 ff. Nagel characterizes the condition as one of "practical solipsism."

and experience may change the order of their desires so that
different ones come uppermost: the prospect of punishment
may dampen an animal's ardor to the point where the animal will refrain from satisfying its appetite, but that is a different matter than calculating what is in your best interests
and being motivated by a conception of your long-term
good. For all of these reasons Veneer Theory seems to me to
be rather silly. I therefore want to set it aside, and talk about
de Waal's more central and interesting question, the question of the roots of morality in our evolved nature, where
they are located and how deep they go.

If someone asked me whether I personally believe that
the other animals are more like human beings than most
people suppose, or whether I believe there is some form of
deep discontinuity between humans and the other animals,
I would have to say yes to both alternatives. In thinking
about this issue it is important to remember that human
beings have a vested interest in what de Waal calls "anthropodenial." We eat nonhuman animals, wear them, perform
painful experiments on them, hold them captive for purposes of our own—sometimes in unhealthy conditions—
we make them work, and we kill them at will. Without even
taking up the urgent moral questions to which these practices give rise, I think it is fair to say that we are more likely
to be comfortable in our treatment of our fellow creatures if
we think that being eaten, worn, experimented on, held
captive, made to work, and killed, cannot mean anything
like the same thing to them that it would to us. And that in
turn seems more likely to the extent they are unlike us in

their emotional and cognitive lives. Of course the fact that we have a vested interest in denying the similarities between ourselves and the other animals does nothing to show that there are such similarities. But once you correct for that vested interest there seems little reason to doubt that observations and experiments of the sort de Waal does and describes, as well as our own everyday interactions with our animal companions, show exactly what they seem to show: that many animals are intelligent, curious, loving, playful, bossy, belligerent creatures in many ways very much like ourselves.

But I don't find a total gradualism very tempting either. To me human beings seem clearly set apart by our elaborate cultures, historical memory, languages with enormously complex grammars and refined expressive power, the practices of art, literature, science, philosophy, and of course of telling jokes. I would also add to this list something that doesn't often appear on it but should—our startling capacity to make friends across the boundaries between species, and to induce the other animals who live with us to do so as well. I am also inclined to agree with Freud and Nietzsche—whose rather gaudier explanations of the evolution of morality don't seem to tempt de Waal very much—that human beings seem psychologically damaged, in ways that suggest some deep break with nature. An old-fashioned philosophical project, dating back to Aristotle, attempts to locate the central difference that accounts for all these other differences between human beings and the other animals. As a very old-fashioned philosopher, I am tempted by that project. What I'd like to do now is talk about one piece of that project that bears on the question of the extent to which morality represents a break with our animal past.

Moral standards are standards governing the way we act, and the question of the extent to which animals are moral or proto-moral beings arises because they unquestionably do act. De Waal's conclusions are largely derived from considering what animals *do*. In his books, de Waal often canvases different possible intentional interpretations of animal behavior and actions, and describes experiments designed to find out which is correct. A capuchin rejects a cucumber when her partner is offered a grape—is she protesting the unfairness, or is she just holding out for a grape? Do the chimps share food because they are grateful to those who have groomed them, or is it just that the grooming has put them in a relaxed and beneficent mood? Sometimes what appear to be evolutionary explanations of animal behavior seem to bleed over into intentional interpretations of their actions, as when de Waal suggests in *Good Natured* that chimpanzees "strive for the kind of community that is in their own best interest."[6] For reasons I have already mentioned, it seems to me difficult to believe a chimpanzee has anything like this on his mind. But in other places de Waal carefully separates the question of the extent to which monkeys and apes do the things he talks about intentionally or deliberately from the question of what explains their tendency to do them. De Waal himself chastises veneer theorists for inferring the selfishness of our intentions from the "selfishness" of our genes.

The question of intention is a question about how an episode in which an animal does something looks from the acting animal's own point of view, whether it is plausible to think that the animal acts with a certain kind of purpose in

---

[6] *Good Natured*, p. 205.

mind. I think there is a temptation to think that the question
whether we can see the origins of morality in animal behavior depends on how exactly we interpret their intentions,
whether their intentions are "good" or not. I think that, at
least taken in the most obvious way, this is a mistake. It
seems to make some sense if you hold the kind of sentimentalist moral theory favored by Hutcheson and Hume, since
according to these thinkers an action gets its moral character
from the fact that onlookers or spectators would approve or
disapprove of it. At least in the case of what Hume called
"the natural virtues," these thinkers believed that the agent
who does a morally good thing need not be motivated by ex-
pressly moral considerations. In fact for this reason, some of
the sentimentalists of the eighteenth century and their critics explicitly discussed the question whether according to
their theories the other animals could be thought of as virtuous. Hutcheson's immediate predecessor, Shaftesbury, had
asserted that you could not count as virtuous unless you
were capable of moral judgment, and that therefore we
would not call a good horse virtuous.[7] But since according
to this sort of theory moral judgment need not play a role in
moral motivation, it is not clear why not. Hutcheson therefore boldly asserted that it is not an absurdity to suppose
that "creatures void of reflection" have some "low virtues."[8]
Although de Waal praises sentimentalist theories, he denies
that his case rests simply on the existence of animals with

---

[7] In *An Inquiry Concerning Virtue or Merit* (1699). I am quoting from D. D.
Raphael's *British Moralists*, vol. I, Indianapolis: Hackett Publishing Company, 1991,
pp. 173–174.

[8] In *An Inquiry Concerning the Original of our Ideas of Virtue or Moral Good*
(1726), *Moralists*, in ibid., vol. I, p. 295. In a later work, Hutcheson argued that it was
confused to think that we can be motivated by moral considerations (*Illustrations on*

intentions we approve of: "whether animals are nice to each other is not the issue, nor does it matter much whether their behavior fits our moral preferences or not. The relevant question rather is whether they possess capacities for reciprocity and revenge, for the enforcement of social rules, for the settlements of disputes, and for sympathy and empathy." (p. 16). But he seems to share an assumption with these early sentimentalists, which is that the morality of an action is a matter of content of the intention with which it is done.

I think this is wrong, and to explain why, I want to take a closer look at the concept of acting intentionally or on purpose. This concept, I believe, does not mark off a single phenomenon, but a number of things that can be ranged on a scale. It is only at a certain point on the scale that the question whether actions have a moral character can arise.

At the bottom of the scale, there is the idea of intentionally or functionally describable movement. The concept of intention in this form applies to any object whatever that has some sort of functional organization, including not only human beings and animals but also plants and machines. Within the economy of a functionally organized object, certain movements can be described as having certain purposes. The heart beats to pump the blood, the alarm rings to wake you up, your computer warns you against a misspelling, the plant's leaves reach out towards the sun to collect its rays. There is no implication that the purposes served

the Moral Sense (1728), ed. Bernard Peach, Cambridge, MA: Harvard University Press, 1971, pp. 139–140). The primary source for Hume's view is Book III of Treatise of Human Nature (1739–1740, 2nd edition, ed. L. A. Selby-Bigge and P. H. Nidditch, Oxford: Oxford University Press, 1978). The primary discussion of the role of moral motivation in moral thought is Book III, Part II, Section One, pp. 477–484.

by these movements are before the minds of the objects that move, or even before the minds of someone who created those objects. Attributing purposes to these movements just reflects the fact that the object is functionally organized.

In the case of living things, especially animals, including the so-called "lower" animals, some of these purposive or intentional movements are guided by the animal's perception. A fish swims upwards towards a surface disturbance that may mean an insect; a cockroach runs under cover as you try to swat him with the newspaper; a spider crawls towards the moth that is caught in the middle of her web. Here we begin to be tempted to use the language of action, and it is clear enough why: when an animal's movements are guided by her perceptions, they are under the control of her mind, and when they are under control of her mind, we are tempted to say that they are under the animal's own control. And this, after all, is what makes the difference between an action and a mere movement—that an action can be attributed to the agent, that it is done under the agent's own control. At this level, should we say that the animal acts intentionally, or on purpose? It depends how you understand the question. The animal is directing her movements and her movement are intentional movements—*the movements* have a purpose. In that sense the animal acts with a purpose, but at this stage there is no need to say that this purpose is somehow before the animal's mind. Admittedly, when we try to look at the situation from the animal's point of view, when we ask ourselves what exactly it is that the animal perceives that determines her movements, it is almost irresistible to describe it purposively. Why does the spider go towards the moth caught in her web unless there is some sense in which the spider sees the moth *as* food and therefore some sense in which she is trying to get food? But however exactly we understand the spider's intention, we need not understand it as a matter of the spider's entertaining thoughts about what she is trying to achieve.

On the other hand, once we are dealing with an intelligent animal, there is no reason *not* to suppose that her purpose is before her mind. Furthermore, I see no reason why we should not suppose that there is a gradual continuum between whatever is going on when a spider's perceptions direct her towards the moth and straightforward cognitive awareness of something as *what you want*. And when such cognitive awareness is in place, presumably the possibility of learning from experience about how to get what you want and avoid what you don't is greatly enhanced. One can always learn from experience by conditioning, but when you are aware of your purpose you can also begin to learn from experience by thinking and remembering.

But even if there is a gradual continuum, it seems right to say that an animal that can entertain his purposes before his mind, and perhaps even entertain thoughts about how to achieve those purposes, is exerting a greater degree of conscious control over his own movements than, say, the spider, and is therefore in a deeper sense an agent. There is now, as in some of de Waal's cases, room for disagreement about what the proper intentional description of an action is, for it is at this level we become committed to keying the intentional description of the action to what is going on from the agent's own point of view. (Freudian slips pose a problem for the claim I just made, but I want to leave that aside for now.) This is a difference from the earlier stage: when we do describe the spider as "trying to get food," we don't care whether that's what the spider thinks she's doing. At the level of the spider, it is natural for the intentional description of the movement and the explanation of it to run together in this way. But once purposes are consciously entertained, the intentional description of the action must capture something about the way it seems to the agent. It's because at this level we key intentional description to the agent's perspective that it makes sense to ask whether the capuchin is protesting the unfairness or merely angling for the grape. So all of this represents a deeper way in which an action may be said to be "intentional."

But some philosophers do not believe that this is the deepest level of intentionality. At the level of intentionality I have just been describing, the animal is aware of his purposes, and thinks about how to pursue them. But he does not choose to pursue those purposes. The animal's purposes are given to him by his affective states: his emotions and his instinctual or learned desires. Even in a case where the animal must choose between two purposes—say a male wants to mate a female but a larger male is coming and he wants to avoid a fight— the choice is made for him by the strength of his affective states. He has learned to fear the larger male more strongly than he desires to mate. The end that the animal pursues is determined for him by his desires and emotions.

Kantians are among the philosophers who believe that a deeper level of assessment and therefore choice is possible. Besides asking yourself how to get what you want most, you can ask yourself whether your wanting this end is a good reason for taking this particular action. The question is not merely about whether the act is an effective way to achieve your end, but whether, even given that it is, your wanting this end *justifies* you in taking this action. Kant of course famously thinks that raising this question about a proposed action takes a particular form: you formulate what he called a maxim—I will do this act in order to achieve this end—and you subject that maxim to the categorical imperative test. You ask whether you can will it as a universal law that everyone who wishes to achieve this sort of end should do this sort of act. In effect you are asking whether your maxim can serve as a rational principle. In some cases, Kant believed, you find you cannot will your maxim as a universal law, and then you have to reject the action described by that maxim as wrong. Even if you do judge the action to be justified and act, you are acting not merely from your desire but from your judgment that the action is justified.

Why do I say this represents a deeper level of intentionality? In the first place, an agent who is capable of this form of assessment is capable of rejecting an action along with its purpose, not because there is something else she wants (or fears) even more, but simply because she judges that doing that sort of act for that purpose is wrong. In a famous passage in the *Critique of Practical Reason*, Kant argued that we are *capable* of setting aside even our most urgent natural desires—the desire to preserve our own lives and to secure the welfare of our loved ones—to avoid performing a wrong action. Kant gives the example of a man who is ordered by his king, on pain of death for himself and suffering for his family, to bear false witness against an innocent person the king wants to get rid of. While no one can say for sure how he would act in such a situation, Kant argues, each of us must admit to himself that he is capable of doing the right thing.[9] Now if we are capable of setting aside our purposes

---

[9] *The Critique of Practical Reason* (1788), translated by Mary Gregor, Cambridge: Cambridge University Press, 1997, p. 27.

when we cannot pursue them by any decent means, then there is also a sense in which when we *do* decide to pursue a purpose, we can be seen as having *adopted* that purpose. Our purposes may be suggested to us by our desires and emotions, but they are not determined for us by our affective states, for if we had judged it wrong to pursue them, we could have laid them aside. Since we choose not only the means to our ends but also the ends themselves, this is intentionality at a deeper level. For we exert a deeper level of control over own movements when we choose our ends as well as the means to them than that exhibited by an animal that pursues ends that are given to her by her affective states, even if she pursues them consciously and intelligently. Another way to put the point is to say that we do not merely *have* intentions, good or bad. We assess and adopt them. We have the capacity for normative self-government, or, as Kant called it, "autonomy." It is at this level that morality emerges. The morality of your action is not a function of the content of your intentions. It is a function of the exercise of normative self-government.[10]

I propose this as an answer to a question de Waal raises in *Good Natured*: "What is different about the way we *act* that makes us, and not any other species, moral beings?" But although I believe the capacity for autonomy is characteristic of human beings and probably unique to human beings, the question how far in the animal kingdom that capacity extends

---

[10] Although it may not seem at all obvious, the argument I have just given is a version of the argument that leads Kant, in the first section of the *Groundwork of the Metaphysics of Morals* (1785), to the conclusion that "an action from duty has its moral worth not in the purpose to be attained by it but in the maxim in accordance with which it is decided upon." I am quoting from the translation by Mary Gregor (Cambridge: Cambridge University Press, 1998), p. 13.

is certainly an empirical one. There is nothing unnatural, nonnatural, or mystical about the capacity for normative self-government. What it requires is a certain form of self-consciousness: namely, consciousness of the grounds on which you propose to act *as grounds*. What I mean is this: a nonhuman agent may be conscious of the object of his fear or desire, and conscious of it as *fearful* or *desirable*, and so as something to be avoided or to be sought. That is the ground of his action. But a rational animal is, in addition, conscious *that* she fears or desires the object, and *that* she is inclined to act in a certain way as a result.[11] That's what I mean by being conscious of the ground *as a ground*. She does not just think about the object that she fears or even about its fearfulness but about her fears and desires themselves. Once you are aware that you are being moved in a certain way, you have a certain reflective distance from the motive, and you are in a position to ask yourself "but should I be moved in that way? Wanting that end inclines me to do that act, but does it really give me a reason to do that act?" You are now in a position to raise a normative question about what you *ought* to do.

I believe that, in general, this form of self-consciousness—consciousness of the grounds of our beliefs and actions—is the source of reason, a capacity that is distinct from intelligence. Intelligence is the ability to learn about the world, to learn from experience, to make new connections of cause and effect, and put that knowledge to work in pursuing your ends. Reason by contrast looks inward, and focuses on the connections between mental states and activities: whether

---

[11]Being conscious of the ground of your beliefs and actions as grounds is a form of *self*-consciousness because it involves identifying *yourself* as the *subject* of certain of your own mental representations.

our actions are justified by our motives or our inferences are justified by our beliefs. I think we could say things about the beliefs of intelligent nonhuman animals that parallel what I am now saying about their actions. Nonhuman animals may have beliefs and may arrive at those beliefs under the influence of evidence, but it is a further step to be the sort of animal that can ask oneself whether the evidence really justifies the belief, and can adjust one's conclusions accordingly.[12]

Both Adam Smith and, following him, Charles Darwin believed that giving an account of the capacity for normative self-government is essential to explaining the development of morality, because it is essential to explaining what Darwin describes as "that short but imperious word *ought*, so full of high significance."[13] And interestingly, both of them explained it by appeal to our social nature.[14] In Smith's account, it is sympathy with the responses of others to ourselves that first turns our attention inward, creating a consciousness of our own motives and characters as objects to be judged. Sympathy, for Smith, is a tendency to put ourselves in the shoes of others and think about the way we would react if we were in their circumstances. We judge another's

[12] I pursue this argument in *The Sources of Normativity*, Cambridge: Cambridge University Press, 1996.

[13] *The Descent of Man*, p. 70.

[14] Freud and Nietzsche also appeal to our social nature to explain the origin of morality. They think that our ability to command ourselves is the result of our internalizing our dominance instincts and turning them against ourselves. Psychologically, the phenomenon of dominance seems to me a promising place to look for the evolutionary origin of the ability to be motivated by an ought, as I proposed in *The Sources of Normativity*, pp. 157–160. For Freud's account, see *Civilization and Its Discontents* (trans. James Strachey, New York: W. W. Norton, 1961), especially chapter VII. For Nietzsche's, see *The Genealogy of Morals* (trans. Walter Kaufman and R. J. Hollingdale, New York: Random House, 1967), especially essay II.

feelings and the resulting actions to be *proper* if they are what we suppose we would feel in his place. If human beings were solitary, Smith argues, our attention would be focused outward: a human afraid of a lion would think about the lion, not about his own fear. Because we are social animals, sympathy leads us to consider how we ourselves appear from the point of view of others, and to enter into their feelings about us. Through the eyes of others we become the spectators of our own conduct, dividing internally, as Smith described it, into an actor and a spectator, and forming judgments about the propriety of our own feelings and motives. The internal spectator transforms our natural desire to be thought well of and praised into something deeper, a desire to be worthy of praise. For to judge that we are worthy of praise is to judge that it would be *proper* for others to praise us, and the internal spectator, who knows our inner motives, is in a position to make a judgment about that. In this way we develop the capacity to be motivated by thoughts about what we ought to do and what we ought to be like.[15]

Darwin speculates that the capacity for normative self-government arose from a difference in the way we are affected by our social instincts and our appetites. The effect of the social instincts on the mind is constant and calm, while that of the appetites is episodic and sharp. Social animals will therefore be under frequent temptations to violate their social instincts for the sake of their appetites, as say when an animal neglects her offspring while mating. But it is a familiar experience that satisfying an appetite seems more important when you are actually in its grip than after you have satisfied it. So

---

[15] Adam Smith, *The Theory of Moral Sentiments* (1759), Indianapolis: Liberty Classics, 1982.

once a social animal's mental faculties develop to the point where she can remember giving way to such temptations, they will seem to her not to have been worth it, and she will eventually learn to control such impulses. Our capacity to be motivated by the imperious word "ought," Darwin suggests, has its origins in this kind of experience.[16]

In an essay called "Conjectures on the Beginnings of Human History," Kant speculated that the form of self-consciousness that underlies our autonomy may also play a role in the explanation of some of the other distinctively human attributes—including culture, romantic love, and the capacity to act from self-interest. Other philosophers have noticed the connection of self-consciousness of this sort with the capacity for language. I can't go into those arguments here, but if they are correct they would provide evidence that only human beings have this form of self-consciousness.[17]

If that is right, then the capacity for normative self-government and the deeper level of intentional control that goes with it is probably unique to human beings. And it is in the proper use of this capacity—the ability to form and act on judgments of what we ought to do—that the essence of morality lies, not in altruism or the pursuit of the greater good. So I do not agree with de Waal when he says, "Instead of merely ameliorating relations around us, as apes do, we

---

[16] *The Descent of Man*, pp. 87–93.

[17] "Conjectures on the Beginning of Human History" (1786) can be found in *Kant: Political Writings*, 2nd ed., ed. Hans Reiss and trans. H. B. Nisbet, Cambridge: Cambridge University Press, 1991.

have explicit teachings about the value of the community and the precedence it takes, or ought to take, over individual interests. Humans go much further in all of this than the apes, which is why we have moral systems and apes do not" (p. 54). The difference here is not a mere matter of degree.

And it isn't a small difference, that ability to be motivated by an ought. It does represent what de Waal calls a saltatory change. A form of life governed by principles and values is a very different thing from a form of life governed by instinct, desire, and emotion—even a very intelligent and sociable form of life governed by instinct, desire, and emotion. Kant's story about the man deciding to face death rather than bear false witness is the stuff of high moral drama, but it has its constant analog in our everyday lives. We have ideas about what we ought to do and to be like and we are constantly trying to live up to them. Apes do not live in that way. We struggle to be honest and courteous and responsible and brave in circumstances where it is difficult. Even if apes are sometimes courteous, responsible, and brave, it is not because they think they should be. Even as primitive a phenomenon as a teenager's efforts to be "cool" is a manifestation of the human tendency to live a life guided by ideals rather than merely driven by impulses and desires. We also suffer deeply from our self-evaluations and act in sick and evil ways as a result. This is part of what I had in mind earlier when I said that human beings seem psychologically damaged in a way that suggests a break with nature. But none of this is a way of saying that morality is a thin veneer on our animal nature. It's the exact contrary: the distinctive character of human action gives us a whole different way of being in the world.

My point is not that human beings live lives of principle and value and so are very noble, while the other animals don't and so are ignoble. The distinctiveness of human action is as much a source of our capacity for evil as of our capacity for good. An animal cannot be judged or held responsible for following its strongest impulse. Animals are not ignoble; they are beyond moral judgment. I agree with de Waal that saying that a person who acts badly acts "like an animal" ("man is wolf to man") can be very misleading in one way. But in another way it is no more an insult to nonhuman animals than saying of a brain-damaged person that he has become a vegetable is an insult to plants. Just as the second remark means that the person has fallen away from his animate nature, the first means that he has fallen away from his human nature. In following his strongest impulse without question or reflection he has failed to exercise his capacity for the kind of intentional control over his movements that makes us human. That is not the only form of wrongdoing, but it is one.

Earlier I said that we are likely to feel more comfortable about the various ways in which we use the other animals if we think they are very different from ourselves. So it is important for me to say that I do not think the difference that I have been describing should provide that comfort. Exactly the opposite is true. In *Good Natured*, de Waal tells a story about an angry capuchin hurling objects at a human observer. When he ran out of other things to throw, the capuchin picked up a squirrel monkey and threw her at the human. De Waal remarks, "Animals often seem to regard those who belong to another kind as merely ambulant objects."[18] But no species is more guilty of treating those who

---

[18] *Good Natured*, p. 84.

belong to other kinds as ambulant objects than we are, and we are the only species that knows it is wrong. As beings who are capable of doing what we ought and holding ourselves responsible for what we do, and as beings who are capable of caring about what we *are* and not just about what we can *get* for ourselves, we are under a strong obligation to treat the other animals decently, even at cost to ourselves.

Ethics and Evolution

How to Get Here from There

PHILIP KITCHER

I

With the possible exception of Jane Goodall, Frans de Waal has done more than any other primatologist to change our understanding of the social lives of our closest living evolutionary relatives. His painstaking observations and experiments have exposed capacities for identifying and responding to the needs of conspecifics, apparently most sophisticated in chimpanzees and bonobos, but present in other primates as well. His detailed accounts of the ways in which these capacities are manifested have broken the stranglehold of the fear, once common among primatologists, that postulating complex psychological states and dispositions is sentimental anthropomorphism. Any scholar who hopes to use primate social behavior as a lens for understanding aspects of our own practices should be profoundly grateful.

In his Tanner Lectures, de Waal intends to build on his decades of careful research to elaborate the program Darwin envisaged in chapter 5 of *The Descent of Man*. Human morality, he suggests, stems from dispositions we share with other primates, particularly with those closest to us on the phylogenetic tree. Yet my formulation of his position, like his own, is vague in crucial respects: what exactly is meant by claiming that morality "stems from" traits present in chimpanzees, or that morality is "a direct outgrowth of the social instincts we share with other animals," or that "deep down" we are truly moral, or that "the building blocks of morality are evolutionarily ancient"? I want to focus the position more precisely by articulating a particular version of what de Waal might have in mind. If this version is not what he intends, I hope it will prompt him to develop his preferred alternative with more specificity than he has done so far.

In fact, I think de Waal's own presentation is hampered by his desire to take a sledgehammer to something he conceives of as the rival to his own view. That rival, "Veneer Theory," is to be demolished. The fact that the demolition is so easy should alert us to the possibility that the real issues have not been exposed and addressed.

II

Veneer Theory, as I understand it, divides the animal kingdom into two. There are nonhuman animals who lack any capacity for sympathy and kindness, and whose actions, to the extent that they can be understood as intentional at all, are the expression of selfish desires. There are also human beings, often driven by selfish impulses to be sure, but capable of rising above egoism to sympathize with others, to curb their baser tendencies, and to sacrifice their own interests for higher ideals. Members of our species have the selfish dispositions that pervade the psychologically more complex parts of the rest of the animal world, but they have something else, an ability to subdue these dispositions. Our psyches are not just full of weeds; we also have a capacity for gardening.

De Waal associates this position with T. H. Huxley, whose famous lecture of 1893 introduced the gardening metaphor. He accuses Huxley of deviating from Darwinism on this point, but it is not clear to me that, even if this is an adequate statement of Huxley's view (which I doubt), the accusation is justified. A fully Darwinian Huxley might claim that human evolution involved the emergence of a psychological trait that has a tendency to inhibit another part of our psychological nature; it is not that something mysterious outside us opposes our nature, but that we come to experience internal conflicts of a kind that had not previously figured in our lives. It would be quite reasonable, of course, to ask this Darwinian Huxley to offer an account of how this new mechanism might have evolved, but, even if any answer proved to be speculative, Huxley would be innocent of assuming that morality is some sort of nonnaturalistic addition.

The version of Veneer Theory I have sketched, and the one that occupies de Waal, takes a specific view of the starting point and the end point. Back in our evolutionary past, we had ancestors, as recent as the common ancestors of human beings and chimpanzees, who lacked any capacities for sympathy and altruism. Present human beings have ways of disciplining their selfish urges, and the theory thinks of morality as this collection of disciplinary strategies. The real objection to Veneer Theory in this form is that it has the starting point wrong. It is falsified by all the evidence de Waal has acquired about the other-directed tendencies of chimpanzees, bonobos, and, to a lesser extent, other primates.

Appreciating this point ought to be the first stage in an inquiry about the evolutionary history that links the psychological dispositions of our ancestors to the capacities that underlie our contemporary moral behavior. De Waal demolishes his favored version of Veneer Theory by being very clear about the starting point—that, after all, is a project to which he has devoted much of his life—but he is considerably less clear as to the nature of the terminus. The vague talk about "building blocks" and "direct outgrowth" comes in because de Waal hasn't thought as hard about the human phenomenon he takes to be anticipated or foreshadowed in chimpanzee social life.

There's a polar opposite of Veneer Theory, one we might call "Solid-to-the-Core Theory" (STCT, for short). STCT claims that morality is essentially present in our evolutionary ancestors. Perhaps in the glory days of human sociobiology some people were tempted to flirt with STCT, supposing, for example, that human morality reduces to dispositions to avoid incest (and similar simple tendencies) and that these have evolutionary explanations that apply to a wide range of organisms.[1] STCT effectively takes the terminus of the

---

[1] See, for example, Michael Ruse and E. O. Wilson, "Moral Philosophy as Applied Science," *Philosophy*, 61, 1986, 173–92. Although this essay takes a radically oversimplified view of the content of morality, I think it would be unfair to accuse Ruse and Wilson of going all the way to STCT. For discussion of the flaws of sociobiological ventures in ethics, see the last chapter of my *Vaulting Ambition* (Cambridge, MA: MIT Press, 1985) and "Four Ways of 'Biologicizing' Ethics" (most easily available in my collection *In Mendel's Mirror* [New York: Oxford University Press, 2003]).

evolutionary process that yields human morality to be the same as some prehuman starting point. It is no more, but no less, plausible than Veneer Theory as de Waal characterizes it. All the interesting positions lie somewhere in between.

De Waal prefaces his lectures with a quotation from the late Stephen Jay Gould, indeed from a passage in which Gould was responding to sociobiological accounts of human nature. I think it's worth reflecting on another observation of Gould's, the comment that when we utter the sentence "Human beings are descended from apes" we can change the emphasis to bring out either the continuities or the differences. Or, to vary the point, Darwin's phrase "descent with modification" captures two aspects of the evolutionary process: descent and modification. What is least satisfactory about de Waal's lectures is his substitution of vague language ("building blocks," "direct outgrowth") for any specific suggestions about what has descended and what has been modified. Lambasting a view like his "Veneer Theory" (or like STCT) is not enough.

III

In fact, de Waal provides a little more than I have so far granted. He has been attuned to developments in evolutionary ethics (or in the evolution of ethics) during the past fifteen years, a period in which the naive reductions favored in sociobiological accounts have given way to proposals of an alliance between Darwin and Hume. The sentimentalist tradition in ethical theory, in which, as de Waal rightly sees, Adam Smith deserves (at least) equal billing with Hume, has won increased favor with philosophers. As it has done so, would-be evolutionary ethicists have felt the appeal of what I shall call the "Hume-Smith lure."

The lure consists in focusing on the central role of sympathy in the ethical accounts offered by Hume and Smith. So you first claim that moral conduct consists in the expression of the appropriate passions, and that sympathy is central to these passions. Then you argue that chimpanzees have capacities for sympathy, and conclude that they have the core of the psychology required for morality. If there are worries about what it means to talk about the "central" role of sympathy or the "core" of moral psychology, the primatologist or evolutionary theorist can shift the burden. Hume, Smith, and their contemporary champions sort out the ways in which sympathy figures in moral psychology and moral behavior; the primatologists demonstrate the sympathetic tendencies at work in primate social life; the evolutionary theorists show how tendencies of this type might have evolved.[2]

My characterization of this strategy as "the Hume-Smith lure" is supposed to signal that it is far more problematic than many writers (including some philosophers, but especially nonphilosophers) take it to be. To understand the difficulties we need to probe the notion of psychological altruism, recognize just what types of psychological altruism have been revealed by studies of primates, and relate these dispositions to the moral sentiments invoked by Hume, Smith, and their successors.

[2] This requires developing the approaches to cooperation pioneered by Robert Trivers, Robert Axelrod, and W. D. Hamilton, so as to take account of the underlying motivations. For one possible approach, see my essay "The Evolution of Human Altruism" (*Journal of Philosophy* 1993; reprinted in *In Mendel's Mirror*).

De Waal wants to recognize nonhuman primates as having dispositions that are not simply egoistic, and it's useful to think of "psychological altruism" as a catchall term for covering these. As I understand it, psychological altruism is a complex notion that involves the adjustment of desires, intentions, and emotions in light of perceptions of the needs and wishes of others. De Waal rightly distinguishes the psychological notion from the biological conception of altruism, defined in terms of the promotion of others' reproductive success at reproductive cost to oneself; as he points out, the interesting notion is one that only applies in the context of intentional behavior, and it can be disconnected from any thought of assisting the reproductive success of other animals.

More precisely, psychological altruism should be thought of in terms of the relation among psychological states in situations that vary according to the perception of another's need or desire. Although an altruistic response can consist in modification of emotions or intentions, it may be easiest to introduce the concept in reference to desire. Imagine an organism A, in a context in which the actions available have no perceptible bearing on another organism B, and suppose that A prefers a particular option. It may nonetheless be true of A that, in a context very similar to the original one, in which there is a perceptible effect on B, A would prefer a different course of action, one that A takes to be more conducive to the wishes or needs of B. If these conditions are met, then A meets a minimal requirement for having an altruistic disposition towards B as a beneficiary. The conditions are not sufficient, however, unless it is also the case that A's change of preference in the situation where B's interests are an issue would be caused by A's perception that the alternative action accorded more closely with B's desires or needs, and, furthermore, that the switch was not generated by a calculation that pursuing the alternative would be likely to satisfy others of A's standing preferences. All this is a way of spelling out the thought that what makes a desire altruistic is a disposition to modify what is chosen in a situation where there is a perceived impact on another, that the modification aligns the choice more closely with the perceived wishes or needs of the other, that the modification is caused by the perception of those wishes or needs, and that it doesn't involve calculation of expected future advantages in satisfaction of standing preferences.

An illustration may help. Suppose that A comes across an item of food, and wishes to devour it all—that is, in the absence of B, A would devour it all. If B is present, however, A may choose to share the food with B (modifying the wish that would have been operative in the context in which B was absent), may do so because A perceives that B desires some of the food (or maybe that B needs some of the food), and may do so not from calculating that sharing will bring some further selfish benefit (for example, that B will then be likely to reciprocate on future occasions). Under these circumstances, A's desire to share is altruistic with respect to B.

We can think of the same structure as applying in the case of emotions or of intentions—a modification of the state that would have been present that is caused by the perception of the wants or needs of the other and that does not come about through the calculation of future benefit. Yet even if we restrict attention to the case of altruistic desire, it should be plain that there are many kinds of psychological altruism. As my disjunctive formulation, "wishes or needs," already suggests, an altruist may respond either to the perceived wants or to the perceived needs of the beneficiary. Typically, these are likely to be in harmony, but, when they diverge, altruists have to choose which to follow. Paternalistic altruism responds to the needs, rather than the wishes; nonpaternalistic altruism does the reverse.

Besides the distinction between paternalistic and nonpaternalistic altruism, it's also important to recognize four dimensions of altruism: intensity, range, extent, and skill. Intensity is marked by the degree to which the altruist accommodates the perceived desire (or need) of the beneficiary; in the food-sharing illustration it's easy to present this concretely, as the fraction of the item the altruist is willing to assign the beneficiary.[3] The range of altruism is marked by the set of contexts in which the altruist makes an altruistic response: to take an example from de Waal, two adult male chimpanzees may be willing to share across a range of situations, but, if the stakes become really high (with the possibility of monopolizing reproductive access, say), an erstwhile friend may act with utter disregard for the other's wishes or needs.[4] The extent of altruism is expressed in the set of individuals towards whom an altruist is prepared to make an altruistic response. Finally the skill of the altruist is measured by the ability to discern, across a range of situations, the real wishes of the intended beneficiary (or, for paternalistic altruists, the real needs of the intended beneficiary).

[3] See "The Evolution of Human Altruism." As noted there, the response can range from complete self-abnegation (give all) through "golden-rule altruism" (split evenly) to complete selfishness (give none).

[4] See Frans de Waal, *Chimpanzee Politics* (Baltimore: Johns Hopkins University Press, 1982).

Even if we ignore the complications of elaborating a similar approach to emotion and intention, and even if we disregard the distinction between paternalistic and nonpaternalistic altruism, it's evident that psychological altruists come in a vast array of types. If we think of a four-dimensional space, we can map "altruism profiles" that capture the distinct intensities and different skills with which individuals respond across a range of contexts and potential beneficiaries. Some possible profiles show low-intensity responses to a lot of others in a lot of situations; other possible profiles show high-intensity responses to a few select individuals across almost all situations; yet others are responses to the neediest individual in any given situation, with the intensity of the response proportioned to the level of need. Which, if any, of these profiles are found in human beings and in nonhuman animals? Which would be found in morally exemplary individuals? Is there a single ideal type to which we'd want everyone to conform, or is a morally ideal world one in which there's diversity?

I pose these questions not as a prelude to answering them, but as a way of exposing how complex the notion of psychological altruism is and how untenable is the idea that, once we know that nonhuman animals have capacities for psychological altruism, we can infer that they have the "building blocks" of morality, too. The demise of Veneer Theory, as de Waal understands it, tells us that our evolutionary relatives belong somewhere in altruism space away from the point of complete selfish indifference. Until we have a clearer view of the specific kinds of psychological altruism chimpanzees (and other nonhuman primates) display, and until we know what kinds are relevant to morality, it's premature to claim that human morality is a "direct outgrowth" of tendencies these animals share.

IV

De Waal has made a powerful case for the existence of *some* forms of psychological altruism in the nonhuman world. His best example, to my mind, is one he offered in *Good Natured*, and which he retells here, the tale of Jakie, Krom, and the tires. His description shows convincingly that the juvenile, Jakie, modified his wishes and intentions from those he'd otherwise have had, that he did so in response to his perception of Krom's wishes, and that the modified wishes were directed at satisfying her perceived desire; although hard-line champions of psychological egoism may insist that the change was produced by some cunning Machiavellian calculation, it's extremely hard to arrive at a plausible hypothesis—Krom is a mildly retarded, low-ranking adult female who is not in any great position to help Jakie, and the idea that this might raise his standing with onlookers is scotched by the absence of other members of the troop.[5] What this reveals is that Jakie was capable of a psychologically altruistic response, of at most moderate intensity (there was little cost in interrupting his activities to help with the

---

[5] It also seems to me that this example avoids the worry that Elliott Sober and David Sloan Wilson belabor in the final chapter of their excellent study of altruism, *Unto Others* (Cambridge, MA: Harvard University Press, 1998). It's very hard to suppose that Jakie was moved by desire for the glow that comes from recognizing that one has acted rightly (or as the community would approve), or by desire to avoid the pang that comes from recognition that one has not. These psychological hypotheses really do invite the charge of unwarranted anthropomorphism.

tires), towards an individual with whom he had a standing relationship, in a context where not much else was going on.

Other examples are a lot less convincing. Consider the capuchins, the cucumber, and the grape. When de Waal's report of his experiments appeared, some enthusiasts were prepared to hail them as demonstrating a sense of fairness in nonhuman animals.[6] I take a sense of fairness to involve psychological altruism, as I have understood it, for it depends on not being content with a situation one would have seen as satisfactory precisely because one recognizes that the needs of others haven't been met. In fact, de Waal's experimental study reveals no kind of psychological altruism, but simply an animal's recognition of the possibility of a preferred reward that it has not received, and a protest that results from the selfish wish for that reward.

In my judgment, the most convincing examples of psychological altruism are those of the Jakie-Krom type, cases in which one animal accommodates its behavior to the perception of a wish, or a need, of another animal with whom it has often interacted, or of instances in which an older animal attends to the perceived needs of the very young. These are quite enough to show that nonhuman animals aren't invariably psychological egoists—and, indeed, to suppose that we are likely to share the same capacities and the same status. But how relevant is psychological altruism of these types to human moral practice?

[6] At a conference at the London School of Economics, de Waal was inclined to present them in similar terms. The Tanner Lectures correctly back away from that interpretation. For, as many people at the LSE meeting pointed out, protests on the part of the aggrieved party don't do much to demonstrate a sense of fairness. Of course, if the lucky capuchin were to throw down the grape until his comrade had a similar reward, that would be *very* interesting!

Some ability to adjust our desires and intentions to the perceived wishes or needs of others appears to be a necessary condition for moral behavior.[7] But, as my remarks about the varieties of psychological altruism should have suggested, it's not sufficient. Hume and Smith both believed that the capacity for psychological altruism, for benevolence (Hume) or sympathy (Smith), was quite limited; Smith begins the *Theory of Moral Sentiments* with a discussion of the ways in which our responses to the emotions of others are pallid copies. Both would probably recognize the full range of de Waal's studies, from *Chimpanzee Politics* through *Peacemaking among Primates* to *Good Natured*, as vindicating their central points, showing (in my terms) that psychological altruism exists, but that it is limited in intensity, range, extent, and skill.

Far more importantly, they would distinguish this first-order psychological altruism from the responses of the genuinely *moral* sentiments. Hume's *Enquiry Concerning the Principles of Morals* closes with the identification of the moral sentiments with "the party of humanity." I interpret him as supposing that we have a capacity for refining the original, limited, dispositions to respond to the wishes and needs of our friends and children. Through proper immersion in society, we can be brought to expand our sympathies, so that we eventually become moved by what is "useful and agreeable" to people, not only when that conflicts with our selfish desires but even when it is at odds with our more primitive, locally partisan, altruistic responses.

---

[7] It seems to me that not only those in the Hume-Smith tradition, but also the strictest Kantians, can accept this point. An extreme Kantian might suppose that the psychologically altruistic response proceeds through the operation of reason, by "cold cognition" rather than by Humean or Smithian sympathy.

Smith is far more explicit than Hume about how this enlargement of sympathy should proceed. He takes it to involve reflecting upon—mirroring—the judgments of those with many different perspectives around us, until we are able to combine each point of view, with its peculiar biases, into an assessment that expresses a genuinely *moral* sentiment.[8] Without the impartial spectator, Smith's "man in the breast," we only have our limited and idiosyncratic sympathies, types of psychological altruism that may be necessary if moral responses are to develop in us but that fall a long way short of morality.

So I think the Hume-Smith lure is just that. It's an invitation to students of animal behavior to demonstrate psychological altruism in their subjects, on the assumption that any kind will do, because "Hume and Smith have shown that altruism is what morality is all about." I think a lot more work needs to be done. Fortunately, de Waal's studies are valuable in showing us how it might proceed.

V

The role of Smith's impartial spectator (or of Kant's inner reasoner, or of a number of other philosophical devices for

---

[8] I describe in more detail how this process of refinement is supposed to occur in "The Hall of Mirrors" (in *Proceedings and Addresses of the American Philosophical Association* November 1985, 67–84.). In that essay, I also argue that Smith's procedure (like Hume's much less developed version) cannot eradicate widely shared biases. Appreciating this point leads me to offer a modification of the ethical project along lines suggested by Dewey—instead of thinking of the enlargement of sympathy as providing a finished ethical system, we should view it as a device for going on from where we are.

directing moral behavior) is especially evident in cases of conflict. The most obvious conflicts are those that pit a selfish impulse against an altruistic one. In these cases, you might think, the verdict of morality is that the altruistic one should win, so that a key step in the evolution of ethics is the acquisition of some capacity for psychological altruism. But that is far too swift. We need impartial spectators (or some equivalent) because our altruistic dispositions are too weak, often of the wrong kinds, and because conflicting altruistic impulses need adjudication.[9] What happens when there's no internal adjudicator can be seen if we consider de Waal's earlier studies in light of his later defenses of psychological altruism.

Chimpanzee Politics and *Peacemaking among Primates* reveal social worlds in which there are limited forms of psychological altruism. The societies are divided into coalitions and alliances, within which, some of the time, the animals cooperate. Some of the cooperation may be based on the identification of future advantages, but there are occasions on which the hypothesis that one animal is responding to the needs of another without calculating future benefit appears quite plausible. If you try to plot the psychological altruism manifested on the dimensions I delineated above, you'll find that de Waal's chimpanzees (the species for which there are most data) are quite limited in the intensity, range, and extent of their altruistic tendencies.

The limitation on range is especially important because, as *Peacemaking among Primates* makes especially vivid, cooperation among these animals, and the psychological altruism

---

[9] Dewey is especially clear on the fact that moral conflict is often not a matter of overcoming the selfish, but deciding which of two conflicting ideals has precedence.

that often underlies it, is always breaking down. When an ally fails to do his part, the social fabric is torn, and has to be repaired. De Waal documents the time-consuming ways in which primates reassure one another, the long bouts of grooming, for example, that follow ruptures within alliances.

If you look at this behavior through the eyes of Adam Smith—both moral philosopher and social theorist—there's an obvious thought. These animals could use their time and energy much more efficiently and profitably than they do, were they to have some device for extending and reinforcing their dispositions to psychological altruism. A "little chimp in the breast" would provide them with a smoother, more functional society, with greater opportunities for cooperative projects; maybe they could even interact with animals whom they didn't see on a daily basis, and their group size could grow. Because they have some forms of psychological altruism they are able to have a richer social organization than most other primate species. Because those forms of psychological altruism are so limited they are socially stuck, unable to achieve larger societies or more extensive cooperation.

Chimpanzee societies show overt conflicts, resolved by elaborate peacemaking. There are also conflicts within the chimpanzees themselves. Sometimes a chimpanzee has a tendency to share that militates against a tendency to keep a food item for itself—the leafy branch is stiffly held towards a beggar and the chimp in possession averts its face, half-turning away;[10] the rigidity of the posture, the redirected gaze, and the expression of discontent make the inner

---

[10] My account here is based on my own, highly limited, observations at the Wild Animal Park near San Diego; the animal I saw belonged to the celebrated bonobo colony there; I don't think that the fact that it was a bonobo, rather than a common chimpanzee, makes any difference to the point.

conflict as clear as in the determined dieter who salivates as she resolutely passes the tempting food tray. The frequency of overt conflicts could be reduced if there were some device for resolving the inner conflicts in the right way. As things stand, however, chimpanzees are wantons (in Harry Frankfurt's helpful terminology), vulnerable to whichever impulse happens to be dominant at a particular moment.

Somewhere in hominid evolution came a step that provided us with a psychological device for overcoming wantonness. I am inclined to think of it as part of what made us fully human. Perhaps it began with an awareness that certain forms of projected behavior might have troublesome results and a consequent ability to inhibit the desires that would otherwise have been dominant. I suspect that it was linked to the evolution of our linguistic capacity, and even that one facet of the selective advantage of linguistic ability lay in helping us to know when to restrain our impulses. As I envisage it, our ancestors became able to formulate patterns for action, to discuss them with one another, and to arrive at ways of regulating the conduct of group members.[11]

At this stage, I conjecture, there began a process of cultural evolution. Different small bands of human beings tried out various sets of normative resources—rules, stories, myths, images, and more—to define the way in which "we" live. Some of these were more popular with neighbors and with descendant groups, perhaps because they offered greater

[11] Here I am indebted to one of the philosophically most sophisticated attempts to set our moral practice in the context of human evolution, Allan Gibbard's *Wise Choices, Apt Feelings* (Cambridge, MA: Harvard University Press, 1990). I think Gibbard is right to emphasize the role of conversation about what to do in the history of moral thinking, from small bands of human beings to the societies of the present.

reproductive success, more likely because they made for smoother societies, greater harmony, and increased cooperation. The most successful ones were transmitted across the generations, appearing in fragmentary ways in the first documents we have, the addenda to law codes of societies in Mesopotamia.

Most of this process is invisible because of the long period between the full acquisition of linguistic ability (50,000 years ago at the very latest) and the invention of writing (5,000 years ago). There are fascinating hints of important developments: the cave art and the figurines, for example. Most significant are the indications of greater ability to cooperate with individuals who don't belong to the local band. From about 20,000 years ago on, the remains of some sites show an increase in the number of individuals present at a particular time, as if several smaller bands had come together there. Even more intriguing are finds of tools made of particular materials at considerable distances from the nearest natural source; perhaps these should be understood in terms of the development of "trading networks," as some archeologists have proposed; or perhaps they should be viewed as indicators of the ability of strangers to negotiate their way through the territories of many different bands. Whichever alternative one selects, these phenomena reveal an increased capacity for cooperation and social interaction, one that becomes fully manifest in the large Neolithic settlements at Jericho and Çatal Hüyük.

Whether or not we can ever do more than guess at the actual course of events, there is, I think, a possible evolutionary account of how we got here from there, one which sees the development of a capacity for normative guidance—perhaps understood in that enlargement and refinement of sympathy that gives rise to Smith's impartial spectator—as a crucial step. Once that was in place, and once we had languages in which to engage in discussions with one another, the explicit moral practices, the compendia of rules, parables, and stories, could be developed in cultural lineages, some of which extend into the present. To revert to Huxley's famous image, we became gardeners, having, as part of our nature, an impulse to root out the weeds that are parts of our psyche, and to foster other plants by adding a stake here or a trellis there. Moreover, with us, as with any garden, the project is never finished but continues indefinitely, as new circumstances and new varieties arise.[12]

VI

In returning to Huxley have I ended up with Veneer Theory? Surely not in the simple version de Waal aims to demolish. How then does it stand with the idea that our evolutionary relatives have the "building blocks" of morality, that our moral practices and dispositions are "direct outgrowths" of capacities we share with them? As I complained earlier, these phrases are too vague to be helpful. There are important continuities between human moral agents and chimpanzees: we share dispositions to psychological altruism without which any genuinely moral action would be impossible. But I suspect that between us and our most recent common ancestor with the chimps there have been some very important evolutionary steps: the emergence of a capacity for normative

---

[12] This is the Deweyan version of the moral project, which I outline more fully in "The Hall of Mirrors."

guidance and self-control, the ability to speak and to discuss potential moral resources with one another, and about fifty thousand years (at least) of important cultural evolution. As Steve Gould saw so clearly, in any evaluation of our evolutionary history you can emphasize the continuities or the discontinuities. I think little is gained by either emphasis. You do better simply to recognize what has endured and what has altered.

Of course, de Waal might reject my speculations about how we got from there to here. Despite the fact that I think my story integrates insights he has developed at different stages of his career, he might prefer some alternative. The important point is that *some* account of this kind is needed. For central to my argument is the thesis that mere demonstration of some type of psychological altruism in chimpanzees (or other higher primates) shows very little about the origins or evolution of ethics. I am happy to consign Veneer Theory (though not Huxley's insights!) to the flames. That, however, is only the start of making the many primatological insights de Waal has given us relevant to our understanding of human morality.

Morality, Reason, and the Rights of Animals

PETER SINGER

My response to Frans de Waal's rich and stimulating Tanner Lectures falls into two parts. The first and longer part raises some issues about the nature of morality, and specifically, de Waal's critique of what he calls "morality as veneer." The second part questions what de-Waal says in his appendix about the moral status of animals. On both these topics I will emphasize points on which I disagree with de Waal, so it is necessary to say here that the positions we share are more important than our differences. I hope that will become apparent in what follows.

DE WAAL'S CRITIQUE OF MORALITY AS VENEER

In *The Expanding Circle*, published in 1981, I argued that the origins of morality are to be found in the nonhuman social mammals from which we evolved. I rejected the view that morality is a matter of culture, rather than biology, or that morality is uniquely human and entirely without roots in our evolutionary history. The development of kin altruism and reciprocal altruism are, I suggested, much more central to our own morality than we recognize.[1] De Waal shares these views, and brings to them a far richer store of knowledge of primate behavior than I could command. It is encouraging to have the support of someone so familiar with our primate relatives, and who affirms, on the basis of that knowledge, the view that there is a great deal of continuity between the social behavior of nonhuman animals and our own moral standards.

De Waal criticizes social contract theory for assuming that there was once a time at which there were humans who were not social beings. Of course, it may be questioned whether any of the major social contract theorists believed they were putting forward a historical account of the origins of morality, but certainly many of their readers have come away with the impression that they did. We may also ask what can be learned from theories that take as their starting point a historically false postulate—that if it were not for the contract, we would be isolated egoists—even if they do not assume this to have actually been the case. Perhaps starting from that point has contributed to what de Waal refers to as the saturation of Western civilization "with the assumption that we are asocial, even nasty creatures."

De Waal rightly rejects the view that all of our morality is "a cultural overlay, a thin veneer hiding an otherwise selfish and brutish nature." Yet because he fails to give sufficient weight to differences he himself acknowledges between primate social behavior and human morality, his dismissal of the Veneer Theory is too swift and he is too harsh with some of its advocates.

---

[1] Peter Singer, *The Expanding Circle*, Oxford: Clarendon Press, 1981.

To understand exactly what de Waal gets right and what he gets wrong, we need to distinguish two different claims:

1. Human nature is inherently social and the roots of human ethics lie in the evolved psychological traits and patterns of behavior that we share with other social mammals, especially primates.
2. All of human ethics derives from our evolved nature as social mammals.

We should accept the first claim and reject the second. But at times de Waal appears to accept them both.

Consider de Waal's critique of T. H. Huxley, who he takes to be the originator of modern Veneer Theory. De Waal writes of "Huxley's curious dualism, which pits morality against nature and humanity against all other animals." As an initial comment, we might note that there is nothing really "curious" about a dualism that has been a standard refrain in one strand—arguably the dominant strand—of Western ethics ever since Plato distinguished different parts of the soul, and likened human nature to a chariot with two horses whom the charioteer must control and make to work together.[2] Kant built the dualism into his metaphysics, suggesting that whereas our desires—including our sympathetic concern for the welfare of others—come from our physical nature, our knowledge of the universal law of morality comes from our nature as rational beings.[3] It's a distinction that has obvious problems, but as we shall see, it would be wrong to dismiss it too lightly.

---

[2] Plato, *The Republic,* especially Books 4, 8, and 9; *Phaedrus,* 246b.

[3] Immanuel Kant, *Groundwork of the Metaphysics of Morals,* trans. Mary Gregor, Cambridge: Cambridge University Press, 1997, sec. III.

Perhaps de Waal thinks that Huxley's position is curious because he is a defender of Darwin, and seems here to be departing from a truly evolutionary approach to ethics. But Darwin himself wrote, in *The Descent of Man*, that "The moral sense perhaps affords the best and highest distinction between man and the lower animals." The differences between Huxley and Darwin on this issue are less clear-cut than de Waal suggests.

That the problem "Veneer Theory" seeks to address is not to be dismissed lightly is perhaps best shown by de Waal's own remarks on Edward Westermarck. De Waal rightly praises Westermarck, whose work is given insufficient attention today. De Waal describes him as "the first scholar to promote an integrated view including both humans and animals and both culture and evolution." Perhaps "the most insightful part of Westermarck's work," in de Waal's opinion, is that in which he tries to distinguish the specifically moral emotions from other emotions. Westermarck, de Waal tells us, "shows that there is more to these emotions than raw gut feeling" and explains that the difference between the moral feelings and "kindred non-moral emotions" is to be found in the "disinterestedness, apparent impartiality, and flavour of generality" shown by the former. De Waal himself elaborates on this thought in the following passage:

> Moral emotions ought to be disconnected from one's immediate situation: they deal with good and bad at a more abstract, disinterested level. It is only when we make general judgments of how anyone ought to be treated that we can begin to speak of moral approval and disapproval. It is in this specific area, famously symbolized by Smith's (1937

[1759]) "impartial spectator," that humans seem to go radically further than other primates.

From where, however, does this concern for judgments made from the perspective of the impartial spectator arise? Not, it appears, from our evolved nature. "Morality likely evolved," de Waal tells us, "as a within-group phenomenon in conjunction with other typical within-group capacities, such as conflict resolution, cooperation, and sharing." Consistently with this idea, de Waal notes that in practice we often fail to put the impartial perspective into practice:

> Universally, humans treat outsiders far worse than members of their own community: in fact, moral rules hardly seem to apply to the outside. True, in modern times there is a movement to expand the circle of morality, and to include even enemy combatants—e.g., the Geneva Convention, adopted in 1949—but we all know how fragile an effort this is.

Consider what de Waal is saying in these passages. On the one hand, we have an evolved nature, which we share with other primates, that gives rise to a morality based on kinship, reciprocity, and empathy with other members of one's own group. On the other hand, the best way of capturing what is distinctive about the moral emotions is that they take an impartial perspective, which leads us to consider the interests of those outside our own group. So central to our current notion of morality is this, that de Waal himself says, as we have just seen, that it is *only* when we make these general, impartial judgments that we can really begin to speak of *moral* approval and disapproval.

The idea of this broadly impartial morality is not new. De Waal quotes Adam Smith, but the idea of a universal morality goes back as least as far as the fifth century before the Christian era, when the Chinese philosopher Mozi, appalled at the damage caused by war, asked: "What is the way of universal love and mutual benefit?" and answered his own question: "It is to regard other people's countries as one's own."[4] Yet, as de Waal points out, the practice of this more impartial morality is "fragile." Doesn't this conception come very close to saying that the impartial element of morality is a veneer, laid over our evolved nature?

In *The Expanding Circle* I suggested that it is our developed capacity to reason that gives us the ability to take the impartial perspective. As reasoning beings, we can abstract from our own case and see that others, outside our group, have interests similar to our own. We can also see that there is no impartial reason why their interests should not count as much as the interests of members of our own group, or indeed as much as our own interests.

Does this mean that the idea of impartial morality is contrary to our evolved nature? Yes, if by "our evolved nature" we mean the nature that we share with the other social mammals from which we evolved. No nonhuman animals, not even the other great apes, come close to matching our capacity to reason. So if this capacity to reason does lie behind the impartial element of our morality, it is something new in evolutionary history. On the other hand, our capacity to reason is part of our nature, and, like every aspect of that nature, is the product of evolution. What makes it different from the other elements of our moral nature is that the evolutionary advantages reason confers are not specifically social. The

[4] Cited from W.-T. Chan, *A Source Book in Chinese Philosophy*, Princeton, NJ: Princeton University Press, 1963, p. 213.

ability to reason gives very general advantages. It does have important social aspects—it helps us to communicate better with others of our species, and hence to cooperate in more detailed plans. But reason also helps us, as individuals, to find food and water, and to understand and avoid threats from predators, or from natural events. It enables us to control fire.

Though a capacity to reason helps us to survive and reproduce, once we develop a capacity for reasoning, we may be led by it to places that are not of any direct advantage to us, in evolutionary terms. Reason is like an escalator—once we step on it, we cannot get off until we have gone where it takes us. An ability to count can be useful, but it leads by a logical process to the abstractions of higher mathematics that have no direct payoff in evolutionary terms. Perhaps the same is true of the capacity to take the perspective of Smith's impartial spectator.[5]

In keeping with this way of looking at the role of reasoning in morality, I differ from de Waal's view of the lessons we should draw from J. D. Greene's innovative work using neuroimaging techniques to help us to understand what happens in moral judgment. De Waal writes:

> Whereas Veneer Theory, with its emphasis on human uniqueness, would predict that moral problem solving is assigned to evolutionarily recent additions to our brain, such as the prefrontal cortex, neuroimaging shows that moral judgment in fact involves a wide variety of brain areas, some extremely ancient (Greene and Haidt 2002). In short, neuroscience seems to be lending support to human morality as evolutionarily anchored in mammalian sociality.

[5] This paragraph draws on Peter Singer, *The Expanding Circle*; see also Colin McGinn, "Evolution, Animals, and the Basis of Morality," *Inquiry* 22 (1979): 91.

To understand why we should not draw this conclusion, we need some more background on what Greene and his colleagues have done. They used neuroimaging to examine brain activity when people respond to situations known in the philosophical literature as "trolley problems."[6] In the standard trolley problem, you are standing by a railroad track when you notice that a trolley, with no one aboard, is rolling down the track, heading for a group of five people. They will all be killed if the trolley continues on its present track. The only thing you can do to prevent these five deaths is to throw a switch that will divert the trolley onto a side track, where it will kill only one person. When asked what you should do in these circumstances, most people say that you should divert the trolley onto the side track, thus saving a net four lives.

In another version of the problem, the trolley, as before, is about to kill five people. This time, however, you are not standing near the track, but on a footbridge above the track. You cannot divert the trolley. You consider jumping off the bridge, in front of the trolley, thus sacrificing yourself to save the imperiled people, but you realize that you are far too light to stop the trolley. Standing next to you, however, is a very large stranger. The only way you can stop the trolley killing five people is by pushing this large stranger off the footbridge, in front of the trolley. If you push the stranger off, he will be killed, but you will save the other five. When

---

[6] Phillipa Foot appears to have been the first philosopher to discuss these problems, in her paper "The Problem of Abortion and the Doctrine of the Double Effect," *Oxford Review* 5 (1967): 5–15; reprinted in James Rachels (ed.), *Moral Problems: A Collection of Philosophical Essays* (New York: Harper & Row, 1971), pp. 28–41. The classic article on the topic, however, is Judith Jarvis Thomson, "Killing, Letting Die, and the Trolley Problem" *The Monist* 59 (1976): 204–217.

asked what you should do in these circumstances, most people say that you should not push the stranger off the bridge.

Greene and his colleagues see these situations as differing in the extent to which they involve an "impersonal" situation such as throwing a switch, or a "personal" violation such as pushing a stranger off a bridge. They found that when subjects were deciding about the "personal" cases, the parts of the brain associated with emotional activity were more active than they were when the subjects were asked to make judgments in "impersonal" cases. More significantly, the minority of subjects who came to the conclusion that it would be right to act in ways that involve a personal violation, but minimize harm overall—for example, those who say that it would be right to push the stranger off the footbridge—show more activity in parts of the brain associated with cognitive activity, and take longer to reach their decision, than those who say "no" to such actions.[7] In other words, when confronted with the need to physically assault another person, our emotions are powerfully aroused, and for some, the fact that this is the only way to save several lives is insufficient to overcome those emotions. But those who are prepared to save as many lives as possible, even if this involves physically pushing another person to his death, appear to be using their reason to override their emotional resistance to the personal violation that pushing another person involves.

Does this lend support for the idea of "human morality as evolutionarily anchored in mammalian sociality"? Yes, to a

---

[7] Joshua Greene and Jonathan Haidt, "How (and Where) Does Moral Judgment Work,?" *Trends in Cognitive Sciences* 6 (2002): 517–523, and personal communications. To be more specific, those who accept the personal violation show more anterior dorsolateral prefrontal activity, while those who reject it have more activity in the precuneus area.

point. The emotional responses that lead most people to say it would be wrong to push a stranger off a footbridge can be explained in just the kind of evolutionary terms that de Waal develops in his lectures, and supports with evidence drawn from his observations of primate behavior. Similarly, it is easy to see why we would not have developed similar responses to something like throwing a switch, which may also cause death or injury, but does so at a distance. For all of our evolutionary history, we have been able to harm people by pushing them violently, but it is only for a few centuries— far too brief a time to make a difference to our evolved nature—that we have been able to harm people by actions like throwing switches.

Before we take this as confirming de Waal's point, however, we need to think again about the subjects of Greene's research who, after some reflection, come to the conclusion that just as it is right to throw a switch to divert a train, killing one person but saving five, so too it is right to push one person off a footbridge, killing one but saving five. This is a judgment that other social mammals seem incapable of making. Yet it too is a moral judgment. It appears to come, not from the common evolutionary heritage we share with other social mammals, but from our capacity to reason. Like the other social mammals, we have automatic, emotional responses to certain kinds of behavior, and these responses constitute a large part of our morality. Unlike the other social mammals, we can reflect on our emotional responses, and choose to reject them. Recall Humphrey Bogart's line in the closing moments of *Casablanca*, when, as Rick Blaine, he tells the woman he loves (Ilsa Lund, played by Ingrid Bergman) to get on the plane and join her husband: "I'm no good at being noble, but it doesn't take much to see that the problems of three little people don't amount to a hill of beans in this crazy world." Maybe it doesn't take much, but it takes capacities that no other social mammals possess.

Although I share de Waal's admiration for David Hume, at this point I find myself developing a reluctant respect for the philosopher who is often seen as Hume's great opponent, Immanuel Kant. Kant thought that morality must be based on reason, not on our desires or emotions.[8] Undoubtedly, he was mistaken to think that morality can be based on reason alone, but it is equally mistaken to see morality only as a matter of emotional or instinctive responses, unchecked by our capacity for critical reasoning. We do not have to accept, as a given, the emotional responses imprinted in our biological nature by millions of years of living in small tribal groups. We are capable of reasoning, and of making choices, and we can reject those emotional responses. Perhaps we do so only on the basis of other emotional responses, but the process involves reason and abstraction, and may lead us, as de Waal acknowledges, to a morality that is more impartial than our evolutionary history as social mammals would—in the absence of that reasoning process—allow.

Just as Kant is not so obviously wrong as de Waal suggests, so too Richard Dawkins has a point when—in a passage that de Waal appears to regard as a lamentable example of Veneer Theory—he writes that "We, alone on earth, can rebel against the tyranny of the selfish replicators."[9] Again, given what de Waal says about the impartial aspect of at least some human morality, it is hard to see why he objects

[8] Immanuel Kant, *Groundwork of the Metaphysics of Morals*, trans. Mary Gregor, sec. II.

[9] Richard Dawkins, *The Selfish Gene*, Oxford: Oxford University Press, 1976, p. 215.

to Dawkins's statement. What Dawkins is saying is not all that different from Darwin's comment, in *The Descent of Man*, that the social instincts "with the aid of active intellectual powers and the effects of habit naturally lead to the golden rule, 'As ye would that men should do to you, do ye to them likewise': and this lies at the foundation of morality."

The issue, then, is not so much whether we accept the Veneer Theory of morality, but rather how much of morality is veneer, and how much is underlying structure. Those who claim that all of morality is a veneer laid over a basically individualistic, selfish human nature, are mistaken. Yet a morality that goes beyond our own group and shows impartial concern for all human beings might well be seen as a veneer over the nature we share with other social mammals.

ANIMAL RIGHTS AND EQUAL CONSIDERATION FOR ANIMALS

In 1993, together with the Italian animal advocate Paola Cavalieri, I cofounded the Great Ape Project, an international effort to gain rights for great apes. The project was simultaneously an idea, an organization, and a book. The book, *The Great Ape Project: Equality beyond Humanity*, includes essays by philosophers, scientists, and experts on the behavior of great apes, including Jane Goodall, Toshisada Nishida, Roger and Deborah Fouts, Lyn White Miles, Francine Patterson, Richard Dawkins, Jared Diamond, and Marc Bekoff. The book begins with a "Declaration on Great Apes" that all the contributors agreed to support. The Declaration demands the extension to all great apes of what it calls "the community of equals," which it defines as "the moral community within which we accept certain basic moral principles or rights as governing our relations with each other and enforceable at law." Among these principles or rights, it asserts, are the right to life, the protection of individual liberty, and the prohibition of torture.

Since the launching of the Great Ape Project, several countries, including Britain, New Zealand, Sweden, and Austria, have banned the use of great apes in medical research. In the United States, though research using chimpanzees continues, it is no longer considered acceptable to kill great apes when their usefulness as experimental subjects is at an end. Instead, they are supposed to be "retired" to sanctuaries, although at present there are not enough sanctuaries to cope with the number of unwanted chimpanzees, and some continue to live in very poor conditions.

My involvement with the Great Ape Project, and perhaps also my long-standing advocacy of "Animal Liberation,"[10] make me, I assume, a target of de Waal's criticism of animal rights advocates in his appendix C. Again, however, it is important to see how much common ground de Waal and I share. He has a strong sense of the reality of animal pain. He firmly rejects those who claim it is "anthropomorphic" to attribute such characteristics as emotions, awareness, understanding, and even politics or culture, to animals. When this rich sense of an animal's subjective experiences is combined with support for "efforts to prevent animal abuse," as it is in de Waal's case, we have come very close to the animal rights position. Once we recognize that nonhuman animals have complex emotional and social needs, we begin to see animal

[10]Peter Singer, *Animal Liberation*, 2nd edition, New York: Ecco, 2003 (first published 1975).

abuse where others might not see it—for example, in the
standard method of keeping pregnant sows in modern in-
tensive farms: on bare concrete, without bedding, isolated in
a metal crate, unable to move freely, to manipulate their environment, to interact with other pigs, or to build a nest in
anticipation of giving birth. If everyone shared de Waal's
views, the animal movement would swiftly achieve its most
important goals.

After agreeing that animals should not be abused, de Waal
adds, "it remains a big leap to say that the only way to insure
their decent treatment is to give them rights and lawyers." I'd
prefer to separate the issues of whether animals should be
granted rights, and whether they should be given lawyers. I
entirely agree with de Waal that people today—and Ameri-
cans in particular—are far too ready to go to court to ad-
vance their aims. The result is a colossal waste of time and
resources, and a tendency for every institution to think de-
fensively about how best to guard itself from a lawsuit. But
recognizing that all animals should have some basic rights
does not necessarily involve bringing in the lawyers. We
could, for example, legislate to protect the rights of animals,
and enforce those laws adequately. Many laws are highly effective because they set standards that virtually everyone is
ready to comply with, without anyone being dragged off to
court. For example, some years ago Britain banned the keeping of sows in the crates described above. As a result, hundreds of thousands of sows have significantly better lives. I
have yet to hear, however, of any British sows having been
given lawyers, or indeed of any need by the authorities to
prosecute farmers for continuing to keep sows in crates after
the prohibition came into effect.

De Waal objects to the idea of animal rights on the

ground that "giving animals rights relies entirely upon our
good will. Consequently, animals will have only those rights
that we can handle. One won't hear much about the rights of
rodents to take over our homes, of starlings to raid cherry
trees, or of dogs to decide their master's walking route.
Rights selectively granted are, in my book, no rights at all."
But giving rights to severely intellectually disabled human
beings also relies entirely on our good will. And all rights are
selectively granted. Babies don't have the right to vote, and
people who, as a result of mental illness or abnormality, have
a tendency to violent antisocial behavior, may lose the right
to liberty. This doesn't mean that the rights to vote, or to liberty, are "no rights at all."

Nevertheless, I don't really disagree with de Waal when he
suggests that instead of talking of the rights of animals, we
could talk of our obligations to them. In the political arena,
claims about rights make wonderful slogans, for they are
rapidly understood to be assertions that someone or some
group is being denied something of importance. It is in that
sense that I support the Declaration on Great Apes, and the
rights for great apes claimed in it. Speaking as a philosopher
rather than an activist, however, whether it is humans or animals who are the subject of our concern, I find claims about
rights unsatisfactory. Different thinkers have produced vary-
ing lists of supposedly self-evident human rights, and arguments for one list rather than another turn rapidly to assertion and counter-assertion. When rights clash, as they
inevitably do, debates about giving one right greater weight
than another usually make little headway. That's because
rights are not really the foundation of our moral obligations. They are themselves based on concern for the interests
of all those affected by our actions—a basic principle that

can be reached by taking the perspective of Smith's "impartial spectator," some refinement on Kant's idea of ensuring that the maxim of your action can be willed as a universal law, or even the more ancient "golden rule."

Taking this perspective of obligation, rather than rights, still requires us to say what weight we will give to the interests of animals. De Waal writes: "we should use the new insights into animals' mental life to foster in humans an ethic of caring in which our interests are not the only ones in the balance." Definitely, we should do at least that. But to acknowledge that human interests are "not the only ones in the balance" is vague. De Waal also writes: "I believe that our first moral obligation is to members of our own species." That is less vague, but it is mere assertion. De Waal does also point out that animal advocates accept medical procedures developed by research on animals, but this is, at best, an *ad hominem* argument against people who may not be morally strong enough to refuse medical assistance when they need it. In fact there are some animal rights advocates who refuse medical treatment developed on animals, although admittedly not many. One might equally well say that we should reject the idea of human equality because one knows of no advocates of this idea who have reduced themselves to penury in order to assist people in other countries who are starving to death. (Again, there are a few—Zell Kravinsky comes very close.[11]) Indeed, the link between the ideal and the suggested action is stronger in the case of human equality and giving to the poor than in the case of animal rights and refusing medical treatment developed through research on animals, because the money we give to the poor would

[11]Ian Parker, "The Gift," *The New Yorker*, August 5, 2004, p. 54.

actually save the lives of some people who, we say, are equal in worth to ourselves, whereas it is not clear how a few people refusing to accept medical treatment would benefit any existing or future animals.

Why should the fact that nonhuman animals are not members of our species justify us in giving less weight to their interests than we give to the similar interests of members of our own species? If we say that moral status depends on membership of our own species, how is our position different from that of the most blatant racists or sexists—those who think that to be white, or male, is to have superior moral status, irrespective of other characteristics or qualities? De Waal finds the animal movement's parallel between the abolition of animal abuse and the abolition of slavery to be "outrageous" because, unlike blacks or women, nonhuman animals can never become full members of our community. That difference does exist, but if animals cannot be full members of our society, neither can humans with severe intellectual disabilities. Yet we don't regard that as a reason for being less concerned about their pain and suffering. In the same way, the fact that animals cannot be full members of our society does not count against giving equal consideration to their interests. If an animal feels pain, the pain matters as much as it does when a human feels pain—if the pain hurts just as much, and will last just as long, and will not have further bad consequences for the human that it does not have for the nonhuman animal. Thus there remains a core of truth in the parallel between human slavery and animal slavery. In both cases, members of a more powerful group arrogate to themselves the right to use beings outside the group for their own selfish purposes, largely ignoring the interests of the outsiders. Then they justify this use by an ideology that explains why members of the more powerful group have superior worth and the right, sometimes god-given, to rule over the outsiders.

Although it is only when animals and humans have similar interests that the principle of equality can straightforwardly be applied—and determining which interests are "similar" is not easy—it is also often difficult to compare different human interests, especially across different cultures. That is no reason for discounting the interests of people with cultures distinct from our own. Granted, the mental capacities of different beings will affect how they experience pain, how they remember it, and whether they anticipate further pain, and these differences can be important. But we would agree that the pain felt by a baby is a bad thing, even if the baby is no more self-aware than, say, a pig, and has no greater capacities for memory or anticipation. Pain can also be a useful warning of danger, so it is not always bad, all things considered. Unless there is some compensating benefit, however, we should consider similar experiences of pain to be equally bad, whatever the species of the being who feels the pain.

Compatibly with this general principle of equal consideration of interests, however, it remains possible to agree with de Waal that "apes deserve special status"—not so much because they are our closest relatives, nor because their similarity to us can "mobilize more guilty feelings about hurting them," but because of what we know about the richness of their emotional and social lives, and their self-awareness and understanding of their situation. Just as such characteristics will often cause humans to suffer more than other animals, so they will often cause great apes to suffer more than mice. But of course, not all research causes suffering, and the test that de Waal thinks research on great apes should pass—that it should be "the sort of research we wouldn't mind doing on human volunteers"—meets the standard of equal consideration of interests.

There is, however, a further reason for giving special status to the great apes. Thanks in part to de Waal's own work, alongside that of Jane Goodall and many others, we know much more about the mental and emotional lives of the great apes than we do about other animals. Because of what we know, and because we can see so much of our own nature in them, the great apes can help to bridge the gulf that millennia of Judeo-Christian indoctrination have dug between us and other animals. Recognizing the great apes as having basic rights would help us to see that the differences between us and other animals are matters of degree, and that could lead to better treatment for all animals.

PART III

RESPONSE TO COMMENTATORS

The Tower of Morality

FRANS DE WAAL

Whereas my respected colleagues focus on what seems missing rather than present in other primates, my own emphasis has rather been on shared characteristics. This reflects my desire to counter the idea that human morality is somehow at odds with our animal background, or even with nature in general. I do appreciate the general support for this position, though, and agree with the repeated suggestions to also consider the discontinuities. So, this is what I intend to do this time around, starting with my definition of morality.

Except, of course, that I would never speak of "discontinuities." Evolution does not occur in leaps: new traits are modifications of old ones so that closely related species differ only gradually. Even if human morality represents a significant step forward, it hardly breaks with the past.

MORAL INCLUSION AND LOYALTY

Morality is a group-oriented phenomenon born from the fact that we rely on a support system for survival (MacIntyre

1999). A solitary person would have no need for morality, nor would a person who lives with others without mutual dependency. Under such circumstances, each individual can just go its own way. There would be no pressure to evolve social constraints or moral tendencies.

In order to promote cooperation and harmony within the community, morality places boundaries on behavior, especially when interests collide. Moral rules create a *modus vivendi* among rich and poor, healthy and sick, old and young, married and unmarried, and so on. Since morality helps people get along and accomplish joint endeavors, it often places the common good above individual interests. It never denies the latter, but insists that we treat others the way we would like to be treated ourselves. More specifically, the moral domain of action is Helping or (not) Hurting others (de Waal 2005). The two H's are interconnected. If you are drowning and I withhold assistance, I am in effect hurting you. The decision to help, or not, is by all accounts a moral one.

Anything unrelated to the two H's falls outside of morality. Those who invoke morality in reference to, say, same-sex marriage or the visibility of a naked breast on prime-time television are merely trying to couch social conventions in moral language. Since social conventions are not necessarily anchored in the needs of others or the community, the harm done by transgressions is often debatable. Social conventions vary greatly: what shocks people in one culture (such as burping after a meal) may be recommended in another. Constrained by their impact on the well-being of others, moral rules are far more constant than social conventions. The golden rule is universal. The moral issues of our time— capital punishment, abortion, euthanasia, and taking care of the old, sick, or poor—all revolve around the eternal themes of life, death, resources, and caring.

Critical resources relating to the two H's are food and mates, which are both subject to rules of possession, division, and exchange. Food is most important for female primates, especially when they are pregnant or lactating (which they are much of the time), and mates are most important for males, whose reproduction depends on the number of fertilized females. This may explain the notorious "double standard" in favor of men when it comes to marital infidelity. Women, on the other hand, tend to be favored in child custody cases, reflecting the primacy assigned to the mother-child bond. Thus, even if we strive for gender-neutral moral standards, real-life judgments are not immune to mammalian biology. A viable moral system rarely lets its rules get out of touch with the biological imperatives of survival and reproduction.

Given how well orientation towards the own group has served humanity for millions of years, and how well it serves us still, a moral system can impossibly give equal consideration to all life on earth. The system has to set priorities. As noted by Pierre-Joseph Proudhon over a century ago: "If everyone is my brother, I have no brothers" (Hardin 1982). On one level, Peter Singer is right to declare all pain in the world equally relevant ("If an animal feels pain, the pain matters as much as it does when a human feels pain"), but on another level, this statement collides head-on with the in-group versus out-group distinction bred in our bones (Berreby 2005). Moral systems are inherently biased towards the in-group.

Morality evolved to deal with the own community first, and has only recently begun to include members of other groups, humanity in general, and nonhuman animals. While

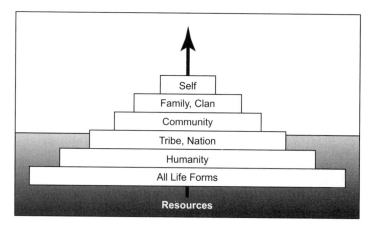

Figure 9 The expanding circle of human morality is actually a floating pyramid viewed from above. Loyalty and duty to immediate family, clan, or species serve as counterforce to moral inclusion. Altruism is spread thinner the further away we get from the center. The pyramid's buoyancy (i.e., available resources) determines how much of it will emerge from the water. The moral inclusion of outer circles is therefore constrained by commitment to inner ones. From de Waal 1996.

applauding the expansion of the circle, this expansion is constrained by affordability, that is, circles are allowed to expand in times of abundance but will inevitably shrink when resources dwindle (figure 9). This is so because the circles track levels of commitment. As stated before: "The circle of morality reaches out farther and farther only if the health and survival of the innermost circles are secure" (de Waal 1996: 213). Since we currently live under affluent circumstances, we can (and ought to) worry about those outside our immediate circle.[1] Nevertheless, a level playing field, in which all circles count equally, clashes with ancient survival strategies.

[1] This view is congruent with Singer's (1972) argument that increased affluence brings increased obligations to those in need.

It is not just that we are biased in favor of the innermost circles (ourselves, our family, our community, our species), we *ought* to be. Loyalty is a moral duty. If I were to come home empty-handed after a day of foraging during a general famine and told my hungry family that I did find bread, yet gave it away, they would be terribly upset. It would be seen as a moral failure, as an injustice, not because the beneficiaries of my behavior did not deserve sustenance, but because of my duty to those close to me. The contrast becomes even starker during war, when solidarity with the own tribe or nation is compulsory: we find treason morally reprehensible.

Animal rights advocates sometimes downplay this tension between loyalty and moral inclusion even though their own behavior tells a different story. When I commented that those who argue against medical research on animals nevertheless make use of it, I was seeking a full acknowledgment that there are two sides to this debate. One cannot silently practice loyalty to the innermost circles (e.g., by accepting for oneself and one's family medical treatments developed on animals) while vocally denying that these circles take priority over other forms of life. Measured along the dimension of kinship, bonding, and group membership, an intellectually disabled human does indeed possess greater moral value than any animal. This loyalty dimension is just as real and important as one that considers sensitivity to pain or self-consciousness. Only by considering both dimensions and reconciling potential conflicts between them can we decide how much moral weight to assign to a sentient being, whether human or animal.

I do worry about animals in medical research, and find it wrenching to decide whether, for example, we should continue hepatitis C research on chimpanzees or forego its potential benefits (compare Gagneux et al. 2005 with Vande-

Berg and Zola 2005). Do we want to cure people or protect chimpanzees? I lean towards protecting chimpanzees in this particular debate, while at the same time admitting that I will take any vaccine that may save my life. The least I can say, therefore, is that I am conflicted. This is why I find animal rights language, with its stridency and absolutes, distinctly unhelpful. It does nothing to lay bare the profound dilemmas that we face. I much prefer a discussion in terms of human *obligations* to animals, especially animals as mentally advanced as the apes, even though I agree with Singer that, in the end, the conclusions we arrive at may not be that different.

THREE LEVELS OF MORALITY

Even if the human moral capacity evolved out of primate group life, this should not be taken to mean that our genes prescribe specific moral solutions. Moral rules are not etched in the genome. An old literature tried to derive the Ten Commandments from the "laws" of biology (e.g., Seton 1907; Lorenz 1974), but such endeavors inevitable fail. Philip Kitcher's tongue-in-cheek Solid-to-the-Core Theory will find few supporters nowadays.

We are not born with any specific moral norms in mind, but with a learning agenda that tells us which information to imbibe. This allows us to figure out, understand, and eventually internalize the moral fabric of our native society (Simon 1990). Because a similar learning agenda underlies language acquisition, I see parallels between the biological foundation of morality and language. In the same way that a child is not born with any particular language, but with the ability to learn *any* language, we are born to absorb moral rules and weigh moral options, making for a thoroughly flexible system that nevertheless revolves around the same two H's and the same basic loyalties it always has.

Level 1: Building Blocks

Human morality can be divided into three distinct levels (table 2), of which the first one-and-a-half seem to have obvious parallels in other primates. Since the upper levels cannot exist without the lower ones, all of human morality is continuous with primate sociality. The first level, extensively discussed in my introductory essay, is the level of moral sentiments, or the psychological "building blocks" of morality. They include empathy and reciprocity, but also retribution, conflict resolution, and a sense of fairness, all of which have been documented in other primates.

In labeling these building blocks, I prefer to employ shared language for humans and apes. Robert Wright's discussion of shared language failed to address the main reason behind it, which is that if two closely related species act similarly the logical default assumption is that the underlying psychology is similar, too (de Waal 1999; appendix A). This holds true regardless of whether we are talking about emotions or cognition, two domains often presented as antithetical even though they are almost impossible to disentangle (Waller 1997). The term "anthropomorphic" is unfortunate as it slaps a disapproving label onto shared language. From an evolutionary perspective, we really have no choice other than to use shared language for similar behavioral phenomena in humans and apes. Most likely, they are *homologous*,

TABLE 2

Three Levels of Morality

Level	Description	Humans and Apes Compared
1. Moral sentiments	Human psychology provides the "building blocks" of morality, such as the capacity for empathy, a tendency for reciprocity, a sense of fairness, and the ability to harmonize relationships.	In all these areas, there exist evident parallels with other primates.
2. Social pressure	Insisting that everyone behaves in a way that favors a cooperative group life. The tools to this end are reward, punishment, and reputation building.	Community concern and prescriptive social rules do exist in other primates, but social pressure is less systematic and less concerned with the goals of society as a whole.
3. Judgment and reasoning	Internalization of others' needs and goals to the degree that these needs and goals figure in our judgment of behavior, including others' behavior that does not directly touch us. Moral judgment is self-reflective (i.e., governs our own behavior as well) and often logically reasoned.	Others' needs and goals may be internalized to some degree, but this is where the similarities end.

that is, derived from shared ancestry. The alternative is to classify similar behavior as *analogous*, that is, independently derived. I realize that social scientists comparing human and animal behavior tend to assume analogy, but with respect to closely related species this assumption strikes the biologist as utterly unparsimonious.

Occasionally, we are able to unravel the mechanisms behind behavior. The example Wright offers of reciprocity based on friendly feelings versus cognitive calculations is a case in point. Over the past twenty years, my coworkers and I have collected systematic data and conducted experiments to illuminate the mechanisms behind observed reciprocity. These mechanisms range from simple to complex. All of the different ways proposed by Wright are actually indicated in other animals. Next to humans, chimpanzees appear to show the cognitively most advanced forms of reciprocity (de Waal 2005; de Waal and Brosnan, 2006).

Level 2: Social Pressure

Whereas the first level of morality seems well developed in our close relatives, at the second level we begin to encounter major differences. This level concerns the social pressure put onto every member of the community to contribute to common goals and uphold agreed-upon social rules. Not that this level is wholly absent in other primates. Chimpanzees do seem to care about the state of affairs within their group and do seem to follow social rules. Recent experiments even indicate conformism (Whiten et al. 2005). But in relation to morality, the most important feature is the already mentioned *community concern* (de Waal 1996), reflected in the way high-ranking females bring conflicted parties together after a fight, thus restoring the peace. Here is the original description of this mediation:

> Especially after serious conflicts between two adult males, the two opponents sometimes were brought together by an adult female. The female approached one of the males, kissed or touched him or presented towards him and then slowly walked towards the other male. If the male followed, he did so very close behind her (often inspecting her genitals) and without looking at the other male. On a few occasions the female looked behind at her follower, and sometimes returned to a male that stayed behind to pull at his arm to make him follow. When the female sat down close to the other male, both males started to groom her and they simply continued grooming after she went off. The only difference being that they groomed each other after this moment, and panted, spluttered, and smacked more frequently and loudly than before the female's departure. (de Waal and van Roosmalen 1979: 62)

Such go-between behavior has been repeatedly observed by my team in a variety of chimpanzee groups. It allows male rivals to approach each other without taking initiative, without making eye contact, and perhaps without losing face. But more importantly: a third party steps in to ameliorate relationships in which she herself is not directly involved.

Policing by high-ranking males shows the same sort of community concern. These males break up fights among others, sometimes standing between them until the conflict calms down. The evenhandedness of male chimpanzees in this role is truly remarkable, as if they place themselves above the contestants. The pacifying effect of this behavior has been documented in both captive (de Waal 1984) and wild chimpanzees (Boehm 1994).[2]

A recent study of policing in macaques has shown that the entire group benefits. In the temporary absence of the usual performers of policing, the remaining group members see their affiliative networks deteriorate and the opportunities for reciprocal exchange dwindle. It is no exaggeration to say, therefore, that in primate groups a few key players can exert extraordinary influence. The group as a whole benefits from their behavior, which enhances social cohesion and cooperation. How and why policing behavior evolved is a separate issue, but its pervasive effect on group dynamics is undeniable (Flack et al. 2005; 2006).

The idea that individuals can make a difference for the group has been taken a giant step further in our own species. We actively insist that each individual try to make a difference for the better. We praise deeds that contribute to the greater good and disapprove of deeds that undermine the social fabric. We approve and disapprove even if our immediate interests are not at stake. I will disapprove of individual A stealing from B not only if I am B, or if I am close to B, but even if I have nothing to do with A and B except for being

---

[2] My popular books do not always present the actual data on which conclusions are based. For example, the claim that high-ranking males police intragroup conflicts in an impartial manner was based on 4,834 interventions analyzed by de Waal (1984). One male, Luit, showed a lack of correlation between his social preferences (measured by association and grooming) and interventions in open conflict. Only Luit showed this dissociation: interventions by other individuals were biased in favor of friends and family. My remark about Luit that "there is no room in this policy for sympathy and antipathy" (de Waal 1998 [1982]: 190) thus summarizes well-quantified aspects of his behavior.

part of the same community. My disapproval reflects concern about what would happen if everyone started acting like A: my long-term interest are not served by rampant stealing. This rather abstract yet still egocentric concern about the quality of life in a community is what underpins the "impartial" and "disinterested" perspective stressed by Philip Kitcher and Peter Singer, which is at the root of our distinction between right and wrong.

Chimpanzees do distinguish between acceptable and unacceptable behavior, but always closely tied to immediate consequences, especially for themselves. Thus, apes and other highly social animals seem capable of developing prescriptive social rules (de Waal 1996; Flack et al. 2004), of which I will offer just one example:

> One balmy evening at the Arnhem Zoo, when the keeper called the chimps inside, two adolescent females refused to enter the building. The weather was superb. They had the whole island to themselves and they loved it. The rule at the zoo was that none of the apes would get fed until all of them had moved inside. The obstinate teenagers caused a grumpy mood among the rest. When they finally did come in, several hours late, they were assigned a separate bedroom by the keeper so as to prevent reprisals. This protected them only temporarily, though. The next morning, out on the island, the entire colony vented its frustration about the delayed meal by a mass pursuit ending in a physical beating of the culprits. That evening, they were the first to come in. (adapted from de Waal 1996: 89)

However impressive such rule enforcement, our species goes considerably further in this than any other. From very young onwards we are subjected to judgments of right and wrong, which become so much part of how we see the world that all behavior shown and all behavior experienced passes through this filter. We put social thumbscrews on everyone, making sure that their behavior fits expectations.[3] We thus build reputations in the eyes of others, who may reward us through so-called "indirect reciprocity" (Trivers 1971; Alexander 1987).

Moral systems thus impose myriad constraints. Behavior that promotes a mutually satisfactory group life is generally considered "right" and behavior that undermines it "wrong." Consistent with the biological imperatives of survival and reproduction, morality strengthens a cooperative society from which everyone benefits and to which most are prepared to contribute. In this sense Rawls (1972) is on target; morality functions as a social contract.

Level 3: Judgment and Reasoning

The third level of morality goes even further, and at this point comparisons with other animals become scarce in-

---

[3] Our experiments on inequity inversion concerned expectations about reward division (Brosnan and de Waal 2003; Brosnan et al. 2005). In response to Philip Kitcher, it should be noted that it is unclear that inequity aversion has much to do with altruism. Another pillar of human morality, equally important as empathy and altruism, is reciprocity and resource distribution. The reactions of primates faced with unequal rewards falls under this domain, showing that they watch what they gain relative to others. Cooperation is not sustainable without a reasonably equal reward distribution (Fehr and Schmidt 1999). Monkeys and apes react negatively to receiving *less* than someone else, which is indeed different from reacting negatively to receiving *more*, but the two reactions may be related if the second reflects anticipation of the first (i.e., if individuals avoid taking more so as to forestall negative reactions in others to such behavior). For a discussion of how these two forms of inequity aversion may relate to the human sense of fairness, see de Waal (2005: 209–11).

deed. Perhaps this reflects just our current state of knowledge, but I know of no parallels in animals for moral reasoning. We, humans, follow an internal compass, judging ourselves (and others) by evaluating the intentions and beliefs that underlie our own (and their) actions. We also look for logic, such as in the above discussion in which moral inclusion based on sentience clashes with moral duties based on ancient loyalties. The desire for an internally consistent moral framework is uniquely human. We are the only ones to worry about why we think what we think. We may wonder, for example, how to reconcile our stance towards abortion with the one towards the death penalty, or under which circumstances stealing may be justifiable. All of this is far more abstract than the concrete behavioral level at which other animals seem to operate.

This is not to say that moral reasoning is totally disconnected from primate social tendencies. I assume that our internal compass is shaped by the social environment. Everyday, we notice the positive or negative reactions to our behavior, and from this experience we derive the goals of others and the needs of our community. We make these goals and needs our own, a process known as *internalization*. Moral norms and values are not argued from independently derived maxims, therefore, but born from internalized interactions with others. A human being growing up in isolation would never arrive at moral reasoning. Such a "Kaspar Hauser" would lack the experience to be sensitive to others' interests, hence lack the ability to look at the world from any perspective other than his or her own. I thus agree with Darwin and Smith (see Christine Korsgaard's commentary) that social interaction must be at the root of moral reasoning.

I consider this level of morality, with its desire for consistency and "disinterestedness," and its careful weighing of what one did against what one could or should have done, uniquely human. Even though it never fully transcends primate social motives (Waller 1997), our internal dialogue nevertheless lifts moral behavior to a level of abstraction and self-reflection unheard of before our species entered the evolutionary scene.

NAILS IN COFFIN

It is good to hear that my "sledgehammer" approach to Veneer Theory (VT) comes down to beating a dead horse (Philip Kitcher) that was silly to begin with (Christine Korsgaard). The only one to have ridden this horse, Robert Wright, now denies having wholeheartedly done so, if at all, whereas Peter Singer defends VT on the grounds that certain aspects of human morality, such as our impartial perspective, appear to be an overlay, hence a sort of veneer.

The latter is quite a different kind of veneer, though. Singer hints at the prominence of layer 3 (judgment and reasoning) in the larger scheme of human morality, but I doubt that he would advocate disconnecting this layer from the other two. This is, however, exactly what VT has tried to achieve by outright denying layer 1 (the moral sentiments) and stressing layer 2 (social pressure) at the expense of everything else. VT presents moral behavior as nothing more than a way of impressing others and building favorable reputations, hence Ghiselin's (1974) equation of an altruist with a hypocrite and Wright's (1994: 344) comment that "To be moral animals, we must realize how thoroughly we aren't." In the words of Korsgaard, VT depicts the human primate as "a creature who lives in a state of deep internal solitude, essentially regarding himself as the only person in a world of potentially useful things—although some of those things have mental and emotional lives and can talk or fight back."

VT occupies an almost autistic universe. One only needs to inspect the indexes of their books to notice that its defenders rarely if ever mention empathy, or other-directed emotions in general. Even though empathy can be overridden by more pressing concerns[4]—which is why universal empathy is such a fragile proposal—its very existence should give pause to anyone depicting us as out only for ourselves. The human tendency to involuntarily flinch at seeing another in pain profoundly contradicts VT's notion of us as self-obsessed. All scientific indications are that we are hardwired to be in tune with the goals and feelings of others, which in turn primes us to take these goals and feelings into account.

Huxley and his followers have tried to drive a wedge between morality and evolution, a position that I attribute to an excessive focus on natural selection. The mistake is to think that a nasty process can only produce nasty outcomes, or as Joyce (2006: 17) recently put it: "the basic blunder [is] confusing the cause of a mental state with its content." Ab-

---

[4] Given a choice between an action that benefits only themselves and an action that benefits both themselves and a companion, chimpanzees seem to make no distinction. Under these circumstances, they only help themselves (Silk et al. 2005). The authors titled their study "Chimpanzees are indifferent to the welfare of unrelated group members," even if all that they demonstrated was that one can create a situation in which chimpanzees consider the welfare of others secondary. I am sure one can do the same with people. When hundreds of people rush into a store that has a rare item for sale, such as a popular Christmas toy, they surely exhibit little regard for the welfare of others. No one, however, would conclude from this that people are incapable of such regard.

sent natural moral inclinations, the only hope VT has for humanity is the semi-religious notion of perfectibility: with great effort we may be able to lift ourselves up by our own bootstraps.[5]

Is VT really too easily countered to be taken seriously, as Philip Kitcher argues? Remember that VT has dominated evolutionary writing for three decades, and lingers still. During this time, anyone who thought differently was labeled "naive," "romantic," "soft-hearted," or worse. I will be more than happy, however, to let VT rest in peace. Maybe the present discussion will serve as the final nail in its coffin. We urgently need to move from a science that stresses narrowly selfish motives to one that considers the self as embedded in and defined by its social environment. This development is well underway in both neuroscience, which increasingly studies shared representations between self and other (e.g., Decety and Chaminade 2003), and economics, which has begun to question the myth of the self-regarding human actor (e.g., Gintis et al. 2005).

FACES OF ALTRUISM

Finally, a few words on selfish versus altruistic motives. This seems like a straightforward distinction, but it is confused by the special way in which biologists employ these terms. First, "selfish" is often a shorthand for self-serving or self-interested. Strictly speaking, this is incorrect, as animals

[5] The idea of a rebellion against base motives, or even against our own genes (Dawkins 1976), is a secular version of the old Christian notion of denial of the flesh. Gray (2002) discusses how religious positions have unconsciously slipped into liberal and scientific discourse.

show a host of self-serving behaviors without the motives and intentions implied by the term "selfish." For example, to say that spiders build webs for selfish reasons is to assume that a spider, while spinning her web, realizes that she is going to catch flies. More than likely, insects are incapable of such predictions. All we can say is that spiders serve their own interests by building webs.

In the same way, the term "altruism" is defined in biology as behavior costly to the performer and beneficial to the recipient regardless of intentions or motives. A bee stinging me when I get too close to her hive is acting altruistically, since the bee will perish (cost) while protecting her hive (benefit). It is unlikely, however, that the bee knowingly sacrifices herself for the hive. The bee's motivational state is hostile rather than altruistic.

So, we need to distinguish intentional selfishness and intentional altruism from mere functional equivalents of such behavior. Biologists use the two almost interchangeably, but Philip Kitcher and Christine Korsgaard are correct to stress the importance of knowing the motives behind behavior. Do animals ever intentionally help each other? Do humans?

I add the second question even if most people blindly assume a affirmative answer. We show a host of behavior, though, for which we develop justifications after the fact. It is entirely possible, in my opinion, that we reach out and touch a grieving family member or lift up a fallen elderly person in the street before we fully realize the consequences of our actions. We are excellent at providing *post hoc* explanations for altruistic impulses. We say such things as "I felt I had to do something," whereas in reality our behavior was automatic and intuitive, following the common human pattern that affect precedes cognition (Zajonc 1980). Similarly, it has been argued that much of our moral decision-making is too rapid to be mediated by the cognition and self-reflection often assumed by moral philosophers (Greene 2005; Kahneman and Sunstein 2005).

We may therefore be less intentionally altruistic than we like to think. While we are *capable* of intentional altruism, we should be open to the possibility that much of the time we arrive at such behavior through rapid-fire psychological processes similar to those of a chimpanzee reaching out to comfort another or sharing food with a beggar. Our vaunted rationality is partly illusory.

Conversely, when considering the altruism of other primates, we need to be clear on what they are likely to know about the consequences of their behavior. For example, the fact that they usually favor kin and reciprocating individuals is hardly an argument against altruistic motives. This argument would only hold if primates consciously considered the return benefits of their behavior, but more than likely they are blind to these. They may evaluate relationships from time to time with respect to mutual benefits, but to believe that a chimpanzee helps another with the explicit purpose of getting help back in the future is to assume a planning capacity for which there is little evidence. And if future payback does not figure in their motivation, their altruism is as genuine as ours (table 3).

If one keeps separate the evolutionary and motivational levels of behavior (known in biology as "ultimate" and "proximate" causes, respectively), it is obvious that animals show altruism at the motivational level. Whether they also do so at the intentional level is harder to determine, since this would require them to know how their behavior impacts the other. Here I agree with Philip Kitcher that the evidence is limited

TABLE 3
Taxonomy of Altruistic Behavior

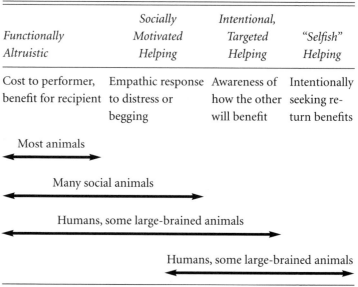

Functionally Altruistic	Socially Motivated Helping	Intentional, Targeted Helping	"Selfish" Helping
Cost to performer, benefit for recipient	Empathic response to distress or begging	Awareness of how the other will benefit	Intentionally seeking return benefits

Most animals

Many social animals

Humans, some large-brained animals

Humans, some large-brained animals

Note: Altruistic behavior falls into four categories dependent on whether or not it is socially motivated and whether or not the actor intends to benefit the other or itself. The vast majority of altruism in the animal kingdom is only functionally altruistic in that it takes place without an appreciation of how the behavior will impact the other and absent any prediction of whether the other will return the service. Social mammals sometimes help others in response to distress or begging (socially motivated helping). Intentional helping may be limited to humans, apes, and a few other large-brained animals. Helping motivated purely by expected return benefits may be rarer still.

even if not wholly absent for large-brained nonhuman mammals, such as apes, dolphins, and elephants, for which we do have accounts of what I call "targeted helping."

Early human societies must have been optimal breeding grounds for survival-of-the-kindest aimed at family and potential reciprocators. Once this sensibility had come into existence, its range expanded. At some point, sympathy for others became a goal in and of itself: the centerpiece of human morality and an essential aspect of religion. It is good to realize, though, that in stressing kindness, our moral systems are enforcing what is already part of our heritage. They are not turning human behavior around, only underlining preexisting capacities.

CONCLUSION

That human morality elaborates upon preexisting tendencies is, of course, the central theme of this volume. The debate with my colleagues made me think of Wilson's (1975: 562) recommendation three decades ago that "the time has come for ethics to be removed temporarily from the hands of philosophers and biologicized." We currently seem in the middle of this process, not by pushing philosophers aside but by including them, so that the evolutionary basis of human morality can be illuminated from a variety of disciplinary angles.

To neglect the common ground with other primates, and to deny the evolutionary roots of human morality, would be like arriving at the top of a tower to declare that the rest of the building is irrelevant, that the precious concept of "tower" ought to be reserved for its summit. While making for good academic fights, semantics are mostly a waste of time. Are animals moral? Let us simply conclude that they occupy several floors of the tower of morality. Rejection of even this modest proposal can only result in an impoverished view of the structure as a whole.

References

Adolphs, R., L. Cahill, R. Schul, and R. Babinsky. 1997. Impaired declarative memory for emotional material following bilateral amygdala damage in humans. *Learning & Memory*, 4: 291–300.

Adolphs, R., H. Damasio, D. Tranel, G. Cooper, and A. R. Damasio. 2000. A role for somatosensory cortices in the visual recognition of emotion as revealed by three-dimensional lesion mapping. *Journal of Neuroscience* 20: 2683–2690.

Adolphs, R., D. Tranel, H. Damasio, and A. R. Damasio. 1994. Impaired recognition of emotion in facial expressions following bilateral damage to the human amygdala. *Nature* 372: 669–672.

Alexander, R. A. 1987. *The Biology of Moral Systems.* New York: Aldine de Gruyter.

Arnhart, L. 1998. *Darwinian Natural Right: The Biological Ethics of Human Nature.* Albany, NY: SUNY Press.

———. 1999. E. O. Wilson has more in common with Thomas Aquinas than he realizes. *Christianity Today International* 5 (6): 36.

Aureli, F., M. Cords, and C. P. van Schaik. 2002. Conflict resolution following aggression in gregarious animals: A predictive framework. *Animal Behaviour* 64: 325–343.

Aureli, F., R. Cozzolino, C. Cordischi, and S. Scucchi. 1992. Kin-oriented redirection among Japanese macaques: An expression of a revenge system? *Animal Behaviour* 44: 283–291.

Aureli, F., and F.B.M. de Waal. 2000. *Natural Conflict Resolution.* Berkeley: University of California Press.

Axelrod, R., and W. D. Hamilton. 1981. The evolution of cooperation. *Science* 211: 1390–1396.

Badcock, C. R. 1986. *The Problem of Altruism: Freudian-Darwinian Solutions.* Oxford: Blackwell.

Bargh, J. A., and T. L. Chartrand. 1999. The Unbearable Automaticity of Being. *American Psychologist* 54: 462–479.

Baron-Cohen, S. 2000. Theory of Mind and autism: A fifteen year review. In *Understanding Other Minds,* ed. S. Baron-Cohen, H. Tager-Flusberg, and D. J. Cohen, pp. 3–20. Oxford: Oxford University Press.

———. 2003. *The Essential Difference.* New York: BasicBooks.

———. 2004. Sex differences in social development: Lessons from autism. In *Social and Moral Development: Emerging Evidence on the Toddler Years,* ed. L. A. Leavitt and D.M.B. Hall, pp. 125–141. Johnson & Johnson Pediatric Institute.

Batson, C. D. 1990. How social an animal? The human capacity for caring. *American Psychologist* 45: 336–346.

Batson, C.D., J. Fultz, and P. A. Schoenrade. 1987. Distress and empathy: Two qualitatively distinct vicarious emotions with different motivational consequences. *Journal of Personality* 55: 19–39.

Berreby, D. 2005. *Us and Them: Understanding Your Tribal Mind.* New York: Little Brown.

Bischof-Köhler, D. 1988. Über den Zusammenhang von Empathie und der Fähigkeit sich im Spiegel zu erkennen. *Schweizerische Zeitschrift für Psychologie* 47: 147–159.

Boehm, C. 1994, Pacifying interventions at Arnhem Zoo and Gombe. In *Chimpanzee Cultures,* ed. R. W. Wrangham, W. C. McGrew, F.B.M. de Waal, and P. G. Heltne, pp. 211–226. Cambridge, MA: Harvard University Press.

———. 1999. *Hierarchy in the Forest: The Evolution of Egalitarian Behavior.* Cambridge, MA: Harvard University Press.

Bonnie, K. E., and F.B.M. de Waal. 2004. Primate social reciprocity and the origin of gratitude. In *The Psychology of Gratitude,* ed. R. A. Emmons and M. E. McCullough, pp. 213–229. Oxford: Oxford University Press.

Bowlby, J. 1958. The nature of the child's tie to his mother. *International Journal of Psycho-Analysis* 39: 350–373.

Bräuer, J., Call, J., and Tomasello, M. 2005. All great ape species follow gaze to distant locations and around barriers. *Journal of Comparative Psychology* 119: 145–154.

Brosnan, S. F., and F.B.M. de Waal. 2003. Monkeys reject unequal pay. *Nature* 425: 297–299.

Brosnan, S. F., Schiff, H., and de Waal, F.B.M. 2005. Tolerance for inequity increases with social closeness in chimpanzees. *Proceedings of the Royal Society* B 272: 253–258.

Burghardt, G. M. 1985. Animal awareness: Current perceptions and historical perspective. *American Psychologist* 40: 905–919.

Byrne, R. W., and A. Whiten. 1988. *Machiavellian Intelligence: Social Expertise and the Evolution of Intellect in Monkeys, Apes, and Humans.* Oxford: Oxford University Press.

Caldwell, M. C., and D. K. Caldwell. 1966. Epimeletic (Care-Giving) behavior in cetacea. In *Whales, Dolphins, and Porpoises*, ed. K. S. Norris, pp. 755–789. Berkeley: University of California Press.

Carr, L., M. Iacoboni, M.-C. Dubeau, J. C. Mazziotta, and G. L. Lenzi. 2003. Neural mechanisms of empathy in humans: A relay from neural systems for imitation to limbic areas. *Proceedings of the National Academy of Sciences* 100: 5497–5502.

Cenami Spada, E. 1997. Amorphism, mechanomorphism, and anthropomorphism. *Anthropomorphism, Anecdotes, and Animals*, ed. R. Mitchell, N. Thompson, and L. Miles, pp. 37–49. Albany, NY: SUNY Press.

Cheney, D. L., and R. M. Seyfarth. 1990. *How Monkeys See the World: Inside the Mind of Another Species.* Chicago: University of Chicago Press.

Church, R. M. 1959. Emotional reactions of rats to the pain of others. *Journal of Comparative and Physiological Psychology* 52: 132–134.

Cohen, S., W. J. Doyle, D. P. Skoner, B. S. Rabin, and J. M. Gwaltney. 1997. Social ties and susceptibility to the Common Cold. *Journal of the American Medical Association* 277: 1940–1944.

Connor, R. C., and K. S. Norris. 1982. Are dolphins reciprocal altruists? *American Naturalist* 119: 358–372.

Damasio, A. 1994. *Descartes' Error: Emotion, Reason, and the Human Brain.* New York: Putnam.

Darwin, C. 1982 [1871]. *The Descent of Man, and Selection in Relation to Sex.* Princeton: Princeton University Press.

Dawkins, R. 1976. *The Selfish Gene.* Oxford: Oxford University Press.

————. 1996. [No title.] *Times Literary Supplement.* November 29: 13.

————. 2003. *A Devil's Chaplain: Reflections on Hope, Lies, Science, and Love.* New York: Houghton Mifflin.

de Gelder, B., J. Snyder, D. Greve, G. Gerard, and N. Hadjikhani. 2004. Fear fosters flight: A mechanism for fear contagion when perceiving emotion expressed by a whole body. *Proceedings from the National Academy of Sciences* 101: 16701–16706.

de Waal, F.B.M. 1984. Sex-differences in the formation of coalitions among chimpanzees. *Ethology & Sociobiology* 5: 239–255.

————. 1989a. Food sharing and reciprocal obligations among chimpanzees. *Journal of Human Evolution* 18: 433–459.

————. 1989b. *Peacemaking among Primates.* Cambridge, MA: Harvard University Press.

————. 1991. Complementary methods and convergent evidence in the study of primate social cognition. *Behaviour* 118: 297–320.

————. 1996. *Good Natured: The Origins of Right and Wrong in Humans and Other Animals.* Cambridge, MA: Harvard University Press.

————. 1997a. *Bonobo: The Forgotten Ape.* Berkeley, CA: University of California Press.

————. 1997b. The Chimpanzee's Service Economy: Food for Grooming. *Evolution & Human Behavior* 18: 375–386.

————. 1998 [1982]. *Chimpanzee Politics: Power and Sex among Apes.* Baltimore, MD: Johns Hopkins University Press.

————. 1999. Anthropomorphism and anthropodenial: Consistency in our thinking about humans and other animals. *Philosophical Topics* 27: 255–280.

————. 2000. Primates—a natural heritage of conflict resolution. *Science* 289: 586–590.

————. 2003. On the possibility of animal empathy. In *Feelings and Emotions: The Amsterdam Symposium,* ed. T. Manstead,

N. Frijda, and A. Fischer, pp. 379–399. Cambridge: Cambridge University Press.

de Waal, F.B.M. 2005. How animals do business. *Scientific American* 292(4): 72–79.

———. 2005. *Our Inner Ape.* New York: Riverhead.

de Waal, F.B.M., and F. Aureli. 1996. Consolation, reconciliation, and a possible cognitive difference between macaque and chimpanzee. In *Reaching into Thought: The Minds of the Great Apes,* ed. A. E. Russon, K. A. Bard, and S. T. Parker, pp. 80–110. Cambridge: Cambridge University Press.

de Waal, F.B.M., and L. M. Luttrell. 1988. Mechanisms of social reciprocity in three primate species: Symmetrical relationship characteristics or cognition? *Ethology & Sociobiology* 9: 101–118.

de Waal, F.B.M., and S. F. Brosnan. 2006. Simple and complex reciprocity in primates. In Cooperation in Primates and Humans: Mechanisms and Evolution, ed. P. M. Kappeler and C. P. van Schaik, pp. 85–105. Berlin: Springer.

de Waal, F.B.M., and A. van Roosmalen. 1979. Reconciliation and consolation among chimpanzees. *Behavioral Ecology & Sociobiology* 5: 55–66.

Decety, J., and T. Chaminade 2003a. Neural correlates of feeling sympathy. *Neuropsychologia* 41: 127–138.

———. 2003b. When the self represents the other: A new cognitive neuroscience view on psychological identification. *Consciousness and Cognition* 12: 577–596.

Desmond, A. 1994. *Huxley: From Devil's Disciple to Evolution's High Priest.* New York: Perseus.

Dewey, J. 1993 [1898]. Evolution and ethics. Reprinted in *Evolutionary Ethics,* ed. M. H. Nitecki and D. V. Nitecki, pp. 95–110. Albany: State University of New York Press.

di Pellegrino, G., L. Fadiga, L. Fogassi, V. Gallese, and G. Rizzolatti. 1992. Understanding motor events: A neurophysiological study. *Experimental Brain Research* 91: 176–180.

Dimberg, U. 1982. Facial reactions to facial expressions. *Psychophysiology* 19: 643–647.

———. 1990. Facial electromyographic reactions and autonomic activity to auditory stimuli. *Biological Psychology* 31: 137–147.

Dimberg, U., M. Thunberg, and K. Elmehed. 2000. Unconscious facial reactions to emotional facial expressions. *Psychological Science* 11: 86–89.

Dugatkin, L. A. 1997. *Cooperation among Animals: An Evolutionary Perspective.* New York: Oxford University Press.

Eibl-Eibesfeldt, I. 1974 [1971]. *Love and Hate.* New York: Schocken Books.

Eisenberg, N. 2000. Empathy and Sympathy. In *Handbook of Emotion,* ed. M. Lewis and J. M. Haviland-Jones, pp. 677–691. 2nd ed. New York: Guilford Press.

Eisenberg, N., and J. Strayer. 1987. *Empathy and Its Development.* New York: Cambridge University Press.

Ekman, P. 1982. *Emotion in the Human Face.* 2nd ed. Cambridge: Cambridge University Press.

Fehr, E., and K. M. Schmidt. 1999. A theory of fairness, competition, and cooperation. *Quarterly Journal of Economics* 114: 817–868.

Feistner, A.T.C., and W. C. McGrew. 1989. Food-sharing in primates: A critical review. In *Perspectives in Primate Biology,* ed. P. K. Seth and S. Seth, vol. 3, pp. 21–36. New Delhi: Today & Tomorrow's Printers and Publishers.

Flack, J. C., and F.B.M. de Waal. 2000. "Any animal whatever": Darwinian building blocks of morality in monkeys and apes. *Journal of Consciousness Studies* 7: 1–29.

Flack, J. C., M. Girvan, F.B.M. de Waal, and D. C. Krakauer. 2006. Policing stabilizes construction of social niches in primates. *Nature* 439: 426–429.

Flack, J. C., L. A. Jeannotte, and F.B.M. de Waal. 2004. Play signaling and the perception of social rules by juvenile chimpanzees. *Journal of Comparative Psychology* 118: 149–159.

Flack, J. C., D. C. Krakauer, and F.B.M. de Waal. 2005. Robustness mechanisms in primate societies: A perturbation study. *Proceedings of the Royal Society London* B 272: 1091–1099.

Frank, R. H. 1988. *Passions within Reason: The Strategic Role of the Emotions.* New York: Norton.

Freud, S. 1962 [1913]. *Totem and Taboo.* New York: Norton.

———. 1961 [1930]. *Civilization and its Discontents.* New York: Norton.

Gagneux, P., J. J. Moore, and A. Varki. 2005. The ethics of research on great apes. *Nature* 437: 27–29.

Gallese, V. 2001. The "shared manifold" hypothesis: From mirror neurons to empathy. In *Between Ourselves: Second-Person Issues in the Study of Consciousness*, ed. E. Thompson, pp. 33–50. Thorverton, UK: Imprint Academic.

Gallup, G. G. 1982. Self-awareness and the emergence of mind in primates. *American Journal of Primatology* 2: 237–248.

Gauthier, D. 1986. *Morals by Agreement*. Oxford: Clarendon Press.

Ghiselin, M. 1974. *The Economy of Nature and the Evolution of Sex*. Berkeley: University of California Press.

Gintis, H., S. Bowles, R. Boyd, and E. Fehr. 2005. *Moral Sentiments and Material Interests*. Cambridge, MA: MIT Press.

Goodall, J. 1990. *Through a Window: My Thirty Years with the Chimpanzees of Gombe*. Boston: Houghton Mifflin.

Gould, S. J. 1980. So cleverly kind an animal. In *Ever Since Darwin*, pp. 260–267. Harmondsworth, UK: Penguin.

Gray, J. 2002. *Straw Dogs: Thoughts on Humans and Other Animals*. London: Granta.

Greene, J. 2005. Emotion and cognition in moral judgment: Evidence from neuroimaging. In *Neurobiology of Human Values*, ed. J.-P. Changeux, A. R. Damasio, W. Singer, and Y. Christen, pp. 57–66. Berlin: Springer.

Greene, J., and J. Haidt. 2002. How (and where) does moral judgement work? *Trends in Cognitive Sciences* 16: 517–523.

Greenspan, S. I., and S. G. Shanker. 2004. *The First Idea*. Cambridge, MA: Da Capo Press.

Haidt, J. 2001. The emotional dog and its rational tail: A social intuitionist approach to moral judgment. *Psychological Review* 108: 814–834.

Hammock, E.A.D., and L. J. Young. 2005. Microsatellite instability generates diversity in brain and sociobehavioral traits. *Science* 308: 1630–1634.

Harcourt, A. H., and F.B.M. de Waal. 1992. *Coalitions and Alliances in Humans and Other Animals*. Oxford: Oxford University Press.

Hardin, G. 1982. Discriminating altruisms. *Zygon* 17: 163–186.

Hare, B., J. Call, and M. Tomasello. 2001. Do chimpanzees know what conspecifics know? *Animal Behaviour* 61, 139–151.

Hare, B., J. Call, and M. Tomasello. In press. Chimpanzees deceive a human competitor by hiding. *Cognition.*

Hare, B., and M. Tomasello. 2004. Chimpanzees are more skilful in competitive than in cooperative cognitive tasks. *Animal Behaviour* 68: 571–581.

Harlow, H. F., and M. K. Harlow. 1965. The affectional systems. In *Behavior of Nonhuman Primates*, ed. A. M. Schrier, H. F. Harlow, and F. Stollnitz, pp. 287–334. New York: Acad. Press.

Hatfield, E., J. T. Cacioppo, and R. L. Rapson. 1993. Emotional Contagion. *Current Directions in Psychological Science* 2: 96–99.

Hauser, M. D. 2000. *Wild Minds: What Animals Really Think.* New York: Holt.

Hebb, D. O. 1946. Emotion in man and animal: An analysis of the intuitive processes of recognition. *Psychological Review* 53: 88–106.

Hediger, H. 1955. *Studies in the Psychology and Behaviour of Animals in Zoos and Circuses.* London: Buttersworth.

Hirata, S. 2006. Tactical deception and understanding of others in chimpanzees. In *Cognitive Development in Chimpanzees*, ed. T. Matsuzawa, M. Tomanaga, and M. Tanaka, pp. 265–276. Tokyo: Springer Verlag.

Hirschleifer, J. 1987. In *The Latest on the Best: Essays in Evolution and Optimality*, ed. J. Dupre, pp. 307–326. Cambridge, MA: MIT Press.

Hobbes, T. 1991 [1651]. *Leviathan.* Cambridge: Cambridge University Press.

Hoffman, M. L. 1975. Developmental synthesis of affect and cognition and its implications for altruistic motivation. *Developmental Psychology* 11: 607–622.

———. 1982. Affect and moral development. *New Directions for Child Development* 16: 83–103.

Hornblow, A. R. 1980. The study of empathy. *New Zealand Psychologist* 9: 19–28.

Hume, D. 1985 [1739]. *A Treatise of Human Nature.* Harmondsworth, UK: Penguin.

Humphrey, N. 1978. Nature's psychologists. *New Scientist* 29: 900–904.

Huxley, T. H. 1989 [1894]. *Evolution and Ethics.* Princeton: Princeton University Press.

Joyce, R. 2006. *The Evolution of Morality.* Cambridge, MA: MIT Press.

Kagan, J. 2000. Human morality is distinctive. *Journal of Consciousness Studies* 7: 46–48.

Kahneman, D., and C. R. Sunstein. 2005. Cognitive psychology and moral intuitions. In *Neurobiology of Human Values,* ed. J.-P. Changeux, A. R. Damasio, W. Singer, and Y. Christen, pp. 91–105. Berlin: Springer.

Katz, L. D. 2000. *Evolutionary Origins of Morality: Cross-Disciplinary Perspectives.* Exeter, UK: Imprint Academic.

Kennedy, J. S. 1992. *The New Anthropomorphism.* Cambridge: Cambridge University Press.

Killen, M., and L. P. Nucci. 1995. Morality, autonomy and social conflict. In *Morality in Everyday Life: Developmental Perspectives,* ed. M. Killen and D. Hart, pp. 52–86. Cambridge: Cambridge University Press.

Kropotkin, P. 1972 [1902]. *Mutual Aid: A Factor of Evolution.* New York: New York University Press.

Kuroshima, H., K. Fujita, I. Adachi, K. Iwata, and A. Fuyuki. 2003. A capuchin monkey (*Cebus apella*) recognizes when people do and do not know the location of food. *Animal Cognition* 6: 283–291.

Ladygina-Kohts, N. N. 2002 [1935]. *Infant Chimpanzee and Human Child: A Classic 1935 Comparative Study of Ape Emotions and Intelligence.* Ed. F.B.M. de Waal. New York: Oxford University Press.

Lipps, T. 1903. Einfühlung, innere Nachahmung und Organempfindung. *Archiv für die gesamte Psychologie* 1: 465–519.

Levenson, R. W., and A. M. Reuf. 1992. Empathy: A physiological substrate. *Journal of Personality and Social Psychology* 63: 234–246.

Lorenz, K. 1974. *Civilized Man's Eight Deadly Sins.* London: Methuen.

Macintyre, A. 1999. *Dependent Rational Animals: Why Human Beings Need the Virtues.* Chicago: Open Court.

MacLean, P. D. 1985. Brain evolution relating to family, play, and the separation call. *Archives of General Psychiatry* 42: 405–417.

Marshall Thomas, E. 1993. *The Hidden Life of Dogs.* Boston: Houghton Mifflin.

Masserman, J., M. S. Wechkin, and W. Terris. 1964. Altruistic Behavior in Rhesus Monkeys. *American Journal of Psychiatry* 121: 584–585.

Mayr, E. 1997. *This Is Biology: The Science of the Living World.* Cambridge, MA: Harvard University Press.

Mencius. n.d. [372–289 BC]. *The Works of Mencius.* English translation by Gu Lu. Shanghai: Shangwu.

Menzel, E. W. 1974. A group of young chimpanzees in a one-acre field. In *Behavior of Non-human Primates,* ed. Schrier, A. M., and Stollnitz, F., vol. 5, pp. 83–153. New York: Academic Press.

Michel, G. F. 1991. Human psychology and the minds of other animals. In *Cognitive Ethology: The Minds of Other Animals,* ed. C. Ristau, pp. 253–272. Hillsdale, NJ: Erlbaum.

Midgley, M. 1979. Gene-Juggling. *Philosophy* 54: 439–458.

Mitchell, R., N. Thompson, and L. Miles. 1997. *Anthropomorphism, Anecdotes, and Animals.* Albany, NY: SUNY Press.

Moss, C. 1988. *Elephant Memories: Thirteen Years in the Life of an Elephant Family.* New York: Fawcett Columbine.

O'Connell, S. M. 1995. Empathy in chimpanzees: Evidence for Theory of Mind? *Primates* 36: 397–410.

Panksepp, J. 1998. *Affective Neuroscience: The Foundations of Human and Animal Emotions.* Oxford: Oxford University Press.

Payne, K. 1998. *Silent Thunder.* New York: Simon & Schuster.

Pinker, S. 1994. *The Language Instinct.* New York: Morrow.

Plomin, R., et al. 1993. Genetic change and continuities from fourteen to twenty months: The MacArthur longitudinal twin study. *Child Development* 64: 1354–1376.

Povinelli, D. J. 1998. Can animals empathize? Maybe not. *Scientific American:* http://geowords.com/lostlinks/b36/7.htm.

———. 2000. *Folk Physics for Apes.* Oxford: Oxford University Press.

Premack, D., and G. Woodruff. 1978. Does the chimpanzee have a theory of mind? *Behavioral & Brain Sciences* 4: 515–526.

Preston, S. D., and F.B.M. de Waal. 2002a. The communication of emotions and the possibility of empathy in animals. In *Altruistic Love: Science, Philosophy, and Religion in Dialogue*, ed. S. G. Post, L. G. Underwood, J. P. Schloss, and W. B. Hurlbut, pp. 284–308. Oxford: Oxford University Press.

———. 2002b. Empathy: Its ultimate and proximate bases. *Behavioral & Brain Sciences* 25: 1–72.

Prinz, W., and B. Hommel. 2002. *Common Mechanisms in Perception and Action*. Oxford: Oxford University Press.

Pusey, A. E., and C. Packer. 1987. Dispersal and Philopatry. In *Primate Societies*, ed. B. B. Smuts et al., pp. 250–266. Chicago: University of Chicago Press.

Rawls, J. 1972. *A Theory of Justice*. Oxford University Press, Oxford.

Reiss, D., and L. Marino. 2001. Mirror self-recognition in the bottlenose dolphin: A case of cognitive convergence. *Proceedings of the National Academy of Science* 98: 5937–5942.

Rimm-Kaufman, S. E., and J. Kagan. 1996. The psychological significance of changes in skin temperature. *Motivation and Emotion* 20: 63–78.

Roes, F. 1997. An interview of Richard Dawkins. *Human Ethology Bulletin* 12 (1): 1–3.

Rothstein, S. I., and R. R. Pierotti. 1988. Distinctions among reciprocal altruism, kin selection, and cooperation and a model for the initial evolution of beneficent behavior. *Ethology & Sociobiology* 9: 189–209.

Sanfey, A. G., J. K. Rilling, J. A. Aronson, L. E. Nystrom, and J. D. Cohen. 2003. The neural basis of economic decision-making in the ultimatum game. *Science* 300: 1755–1758.

Schleidt, W. M., and M. D. Shalter. 2003. Co-evolution of humans and canids, an alternative view of dog domestication: *Homo homini lupus? Evolution and Cognition* 9: 57–72.

Seton, E. T. 1907. *The Natural History of the Ten Commandments*. New York: Scribner.

Shettleworth, S. J. 1998. *Cognition, Evolution, and Behavior*. New York: Oxford University Press

Shillito, D. J., R. W. Shumaker, G. G. Gallup, and B. B. Beck. 2005. Understanding visual barriers: Evidence for Level 1 perspective taking in an orang-utan, *Pongo pygmaeus. Animal Behaviour* 69: 679–687.

Silk, J. B., S. C. Alberts, and J. Altmann. 2003. Social bonds of female baboons enhance infant survival. *Science* 302: 1231–1234.

Silk, J. B., S. F. Brosnan, Vonk, J., Henrich, J., Povinelli, D. J., Richardson, A. S., Lambeth, S. P., Mascaro, J., and Schapiro, S. J. 2005. Chimpanzees are indifferent to the welfare of unrelated group members. *Nature* 437: 1357–1359.

Simon, H. A. 1990. A mechanism for social selection and successful altruism. *Science* 250: 1665–1668.

Singer, P. 1972. Famine, affluence and morality. *Philosophy & Public Affairs* 1: 229–243.

Singer, T., B. Seymour, J. O'Doherty, K. Holger, R. J. Dolan, and C. D. Frith. 2004. Empathy for pain involves the affective but not sensory components of pain. *Science* 303: 1157–1162.

Smith, A. 1937 [1759]. *A Theory of Moral Sentiments.* New York: Modern Library.

Sober, E. 1990. Let's razor Ockham's Razor. In *Explanation and Its Limits*, ed. D. Knowles, pp. 73–94. Royal Institute of Philosophy Suppl. vol. 27, Cambridge University Press.

Sober, E., and D. S. Wilson. 1998. *Unto Others: The Evolution and Psychology of Unselfish Behavior.* Cambridge, MA: Harvard University Press.

Taylor, C. E., and M. T. McGuire. 1988. Reciprocal altruism: Fifteen years later. *Ethology & Sociobiology* 9: 67–72.

Taylor, S. 2002. *The Tending Instinct.* New York: Times Books.

Todes, D. 1989. *Darwin without Malthus: The Struggle for Existence in Russian Evolutionary Thought.* New York: Oxford University Press.

Tomasello, M. 1999. *The Cultural Origins of Human Cognition.* Cambridge, MA: Harvard University Press.

Tomita, H., M. Ohbayashi, K. Nakahara, I. Hasegawa, and Y. Miyashita. 1999. Top-down signal from prefrontal cortex in executive control of memory retrieval. *Nature* 401: 699–703.

Trevarthen, C. 1993. The function of emotions in early infant communication and development. In *New Perspectives in Early*

Communicative Development, ed. J. Nadel and L. Camaioni, pp. 48–81. London: Routledge.

Trivers, R. L. 1971. The evolution of reciprocal altruism. *Quarterly Review of Biology* 46: 35–57.

VandeBerg, J. L., and S. M. Zola. 2005. A unique biomedical resource at risk. *Nature* 437: 30–32.

van Hooff, J.A.R.A.M. 1967. The facial displays of the Catarrhine monkeys and apes. In *Primate Ethology*, ed. D. Morris, pp. 7–68. Chicago: Aldine.

van Schaik, C. P. 1983. Why are diurnal primates living in groups? *Behaviour* 87: 120–44.

von Uexküll, J. 1909. *Umwelt und Innenwelt der Tiere*. Berlin: Springer.

Waller, B. N. 1997. What rationality adds to animal morality. *Biology & Philosophy* 12: 341–356.

Warneken, F., and M. Tomasello. 2006. Altruistic helping in human infants and young chimpanzees. *Science* 311: 1301–1303.

Watson, J. B. 1930. *Behaviorism*. Chicago: University of Chicago Press.

Watts, D. P., F. Colmenares, and K. Arnold. 2000. Redirection, consolation, and male policing: How targets of aggression interact with bystanders. In *Natural Conflict Resolution*, ed. F. Aureli and F.B.M. de Waal, pp. 281–301. Berkeley: University of California Press.

Wechkin, S., J. H. Masserman, and W. Terris. 1964. Shock to a conspecific as an aversive stimulus. *Psychonomic Science* 1: 47–48.

Westermarck, E. 1912 [1908]. *The Origin and Development of the Moral Ideas*, vol. 1. 2nd ed. London: Macmillan.

———. 1917 [1908]. *The Origin and Development of the Moral Ideas*, vol. 2. 2nd ed. London: Macmillan.

Whiten, A., V. Horner, and F.B.M. de Waal. 2005. Conformity to cultural norms of tool use in chimpanzees. *Nature* 437: 737–740.

Wicker, B., C. Keysers, J. Plailly, J. P. Royet, V. Gallese, and G. Rizzolatti. 2003. Both of us disgusted in my insula: The common neural basis of seeing and feeling disgust. *Neuron* 40: 655–664.

Williams, G. C. 1988. Reply to comments on "Huxley's Evolution and Ethics in Sociobiological Perspective." *Zygon* 23: 437–438.

Williams, J.H.G., A. Whiten, T. Suddendorf, and D. I. Perrett. 2001.

Imitation, mirror neurons and autism. *Neuroscience and Biobehavioral Reviews* 25: 287–295.

Wilson, E. O. 1975. *Sociobiology: The New Synthesis.* Cambridge, MA: Harvard University Press.

Wispé, L. 1991. *The Psychology of Sympathy.* New York: Plenum.

Wolpert, D. M., Z. Ghahramani, and J. R. Flanagan. 2001. Perspectives and problems in motor learning. *Trends in Cognitive Sciences* 5: 487–494.

Wrangham, R. W. 1980. An ecological model of female-bonded primate groups. *Behaviour* 75: 262–300.

Wrangham, R. W., and D. Peterson. 1996. *Demonic Males: Apes and the Evolution of Human Aggression.* Boston: Houghton Mifflin.

Wright, R. 1994. *The Moral Animal: The New Science of Evolutionary Psychology.* New York: Pantheon.

Yerkes, R. M. 1925. *Almost Human.* New York: Century.

Zahn-Waxler, C., B. Hollenbeck, and M. Radke-Yarrow. 1984. The origins of empathy and altruism. In *Advances in Animal Welfare Science,* ed. M. W. Fox and L. D. Mickley, pp. 21–39. Washington, DC: Humane Society of the United States.

Zahn-Waxler, C., and M. Radke-Yarrow. 1990. The origins of empathic concern. *Motivation and Emotion* 14: 107–130.

Zahn-Waxler, C., M. Radke-Yarrow, E. Wagner, and M. Chapman. 1992. Development of concern for others. *Developmental Psychology* 28: 126–136.

Zajonc, R. B. 1980. Feeling and thinking: Preferences need no inferences. *American Psychologist* 35: 151–175.

———. 1984. On the primacy of affect. *American Psychologist* 39: 117–123.

Contributors

Frans de Waal is a Dutch-born ethologist/biologist known for his work on the social intelligence of primates. His first book, *Chimpanzee Politics* (1982), compared the schmoozing and scheming of chimpanzees involved in power struggles with that of human politicians. Ever since, de Waal has drawn parallels between primate and human behavior, from peacemaking and morality to culture. His scientific work has been published in hundreds of technical articles in journals such as *Science, Nature, Scientific American,* and outlets specialized in animal behavior. De Waal is also editor or coeditor of nine scientific volumes. His seven popular books—translated into more than a dozen languages—have made him one of the world's most visible primatologists. His latest is *Our Inner Ape* (2005), published by Riverhead. De Waal is C. H. Candler Professor in the Psychology Department of Emory University and director of the Living Links Center at the Yerkes National Primate Center, in Atlanta, Georgia. He has been elected to the National Academy of Sciences (U.S.) and the Royal Dutch Academy of Sciences.

Philip Kitcher is John Dewey Professor of Philosophy at Columbia University. He is the author of nine books, including, most recently, *In Mendel's Mirror: Philosophical Reflections on Biology* (Oxford, 2003); *Finding an Ending: Reflections on Wagner's Ring* (coauthored with Richard Schacht, Oxford, 2004), and *Life without God: Darwin, Design, and the Future of Faith* (forthcoming from Oxford University Press). He is a past president of the American

Philosophical Association (Pacific Division) and a former editor-in-chief of the journal *Philosophy of Science*. He is a fellow of the American Academy of Arts and Sciences.

Christine M. Korsgaard received her B.A. from the University of Illinois and her Ph.D. from Harvard, where she studied with John Rawls. She taught at Yale, the University of California at Santa Barbara, and the University of Chicago before taking up her present position at Harvard, where she is Arthur Kingsley Porter Professor of Philosophy. She is the author of two books. *Creating the Kingdom of Ends* (Cambridge, 1996) is a collection of previously published essays on Kant's moral philosophy. *The Sources of Normativity* (Cambridge, 1996), an exploration of modern views about the basis of obligation, is an expanded version of her 1992 Tanner Lectures on Human Values. She is currently working on a book on the connections between the metaphysics of agency, the normative standards that govern action, and the constitution of personal identity, entitled *Self-Constitution: Agency, Identity, and Integrity*; and putting together another collection of papers, under the title *The Constitution of Agency: Essays on Practical Reason and Moral Psychology* (both to be published by Oxford).

Stephen Macedo writes and teaches on political theory, ethics, American constitutionalism, and public policy, with an emphasis on liberalism, justice, and the roles of schools, civil society, and public policy in promoting citizenship. He served as founding director of Princeton's Program in Law and Public Affairs (1999–2001). He recently served as vice president of the American Political Science Association and chair of its first standing committee on Civic Education and Engagement, and in this capacity he is principal coauthor of *Democracy at Risk: How Political Choices Undermine Citizenship and What We Can Do About It* (2005). His books include *Diversity and Distrust: Civic Education in a Multicultural Democracy* (2000); and *Liberal Virtues: Citizenship, Virtue, and Community in Liberal Constitutionalism* (1990). He is coauthor and coeditor of *American Constitutional Interpretation*, 3rd edition, with W. F. Murphy, J. E. Fleming, and S. A. Barber. Among his edited volumes are *Educating Citizens: International Perspectives on*

Civic Values and School Choice (2004) and *Universal Jurisdiction: International Courts and the Prosecution of Serious Crimes under International Law* (2004). Macedo has taught at Harvard University and at the Maxwell School at Syracuse University. He earned his B.A. at the College of William and Mary, masters degrees at the London School of Economics and Oxford University, and his M.A. and Ph.D. at Princeton University.

Josiah Ober, formerly the David Magie '97 Class of 1897 Professor of Classics at Princeton University, is the Constantine Mitsotakis Professor of Political Science and Classics at Stanford University. His collected essays *Athenian Legacies: Essays on the Politics of Going on Together* were published by the Princeton University Press in 2005. In addition to his ongoing work on knowledge and innovation in democratic Athens, he is interested in the relationship between democracy as a natural human capacity and its association with moral responsibility.

Peter Singer was educated at the University of Melbourne and the University of Oxford. In 1977, he was appointed to a chair of philosophy at Monash University in Melbourne and subsequently was the founding director of that university's Centre for Human Bioethics. In 1999 he became the Ira W. DeCamp Professor of Bioethics. Peter Singer was the founding president of the International Association of Bioethics, and with Helga Kuhse, founding coeditor of the journal *Bioethics*. He first became well known internationally after the publication of *Animal Liberation*. His other books include: *Democracy and Disobedience; Practical Ethics; The Expanding Circle; Marx; Hegel; The Reproduction Revolution* (with Deane Wells), *Should the Baby Live?* (with Helga Kuhse), *How Are We to Live?; Rethinking Life and Death; One World; Pushing Time Away;* and *The President of Good and Evil*. His works have appeared in twenty languages. He is the author of the major article on ethics in the current edition of the Encyclopaedia Britannica.

Robert Wright is the author of *Nonzero: The Logic of Human Destiny* and *The Moral Animal: Evolutionary Psychology and Everyday Life*, both published by Vintage Books. *The Moral Animal* was named by the *New York Times Book Review* as one of the twelve best books of 1994 and has been published in twelve languages. *Nonzero* was named a *New York Times Book Review* Notable Book for 2000 and has been published in nine languages. Wright's first book, *Three Scientists and Their Gods: Looking for Meaning in an Age of Information*, was published in 1988 and was nominated for a National Book Critics Circle Award. Wright is a contributing editor at the *New Republic*, *Time*, and *Slate*. He has also written for the *Atlantic Monthly*, the *New Yorker*, and the *New York Times Magazine*. He previously worked at *The Sciences* magazine, where his column "The Information Age" won the National Magazine Award for Essay and Criticism.

Index

altruism: cognitive vs. emotional motivations for reciprocal, 85–88; definition of in biology, 178; dimensions of, 128–29; empathy and sympathy, relation between and, 28; evolution of as central to human morality, 141; examples of primate, 29–33; the expanding circle of morality and, 164–65; helping tendencies, reciprocal as alternative to group selection in explaining, 15–16; paternalistic and nonpaternalistic, distinction between, 128; psychological (*see* psychological altruism); retributive kindly emotions as parallel to reciprocal, 19–20; selfishness vs., intentionality in distinguishing, 177–81; taxonomy of, 180. *See also* sympathy

animal rights: the Great Ape Project, 151–52, 154; human obligations to nonhuman animals, 118–19, 155–58, 166; medical research and (*see* medical research); responses to skepticism regarding, 153–54; skepticism regarding, 75–77, 154–55, 165–66

animals, nonhuman. *See* nonhuman animals

animal testing. *See* medical research

anthropocentrism, xvii

anthropodenial, xvi, 65, 67, 103

anthropomorphic parsimony, principle of, 92–93

anthropomorphism: chimpanzees, language of appropriate for, 83–96; cognitive *vs.* evolutionary parsimony and, 61–63; the debate regarding, xvi–xvii; definitions of, 63; the dilemma regarding, 59–67; emotional vs. cognitive language of, 84–89; emotional vs. cognitive language of, preferences for, 89–92, 95–96; fear of, stifling of research into animal emotions due to, 25; labeling shared language as, 167; scientific distinguished from sentimental, xvi; unitary explanation of shared characteristics vs. anthropodenial, 65–66

apes: bonobos, 71–73; chimpanzees (*see* chimpanzees); humans and, comparison of regarding levels of morality, 168; medical research, use of for, 78–80; special status for, 78–79, 157–58; theory of mind in, 69–73. *See also* primates

reciprocity: among chimpanzees, 42–44, 169; as building block of morality, 20–21; as building block of naturalistic theory of human morality, 53; definition of, 13; fairness and, 42–49; the Golden Rule and human morality, at the heart of, 49; indirect, 20, 173
reconciliation, 19
religion, x–xi, 177n
retributive emotions, 18–20, 44
rhesus monkeys: emotional contagion among infant, 27; sympathy in, 29. *See also* primates
Ruse, Michael, 123n
Russian dolls, xiv, 21, 39–40

scientific anthropomorphism: distinguished from sentimental anthropomorphism, xvi. *See also* anthropomorphism
selection, evolutionary process of. *See* natural selection
self-consciousness, 36, 113–17
self-deception, 11
selfishness/self-interest: altruism vs., intentionality in distinguishing, 177–80; nonhuman animals and, 102–3; presence of, overestimating the, 52; usage of the term(s), 13–14, 177–78; Veneer Theory and, xi, 11–12, 99–103, 121–22
Seyfarth, R. M., 65
Shaftesbury, Earl of. *See* Cooper, Anthony Ashley
Shanker, S. G., 23
Sidgwick, Henry, 100
Singer, Peter: affluence increases obligation, 164n; all pain is equally relevant, 163; conclusions of, similarity

to de Waal's, 166; de Waal's circle of morality and the extension of morality to animals, xv–xvi; impartial/disinterested perspective, significance of, 172; trolley problems, xviii; Veneer Theory, limited defense of, 175
Smith, Adam: empathic capacity, description of, 31; impartial spectator of, 20, 144, 146, 155; self-government, moral significance of the capacity for, 114–15; on sympathy, 15, 51, 124–25, 132–33
Sober, Elliott, 11–12, 130n
social contract theory, xi, 3–5, 141
sociality, human. *See* humans/human nature
social pressure, 169–73
Solid-to-the-Core Theory (STCT), 123–24, 166
somatic marker hypothesis, 38
spectatorism, xvii
speech. *See* language
STCT. *See* Solid-to-the-Core Theory
sympathetic resentment, 19
sympathy: in chimpanzees, 30; Darwin on, 14–15; definition of, 26–27; empathy as leading to, xiii–xiv; as a natural, involuntary emotion, 51; in nonhuman animals, ignoring of, 24–25; in sentimentalist moral theory, 124–25, 132–33; Smith on, 15, 114–15. *See also* empathy

targeted helping, 32, 36, 40–41
Theory of Mind (ToM), xvi–xvii, 31, 37, 69–73
Thomas, Marshall, 64
ToM. *See* Theory of Mind
Trivers, Robert L., 12, 44, 125n
trolley problems, xviii, 147–49